Frederick Guttmann R.

TREATISE ON THE CREATION OF THE COSMOS

First edition. April 11, 2024.

ISBN: 979-8231854271

Written by Frederick Guttmann.

TREATISE ON THE CREATION OF THE COSMOS

Frederick Guttmann R.

Project Magen (June - 2017)
www.frederickguttmann.com[1]
87 pages

1. *http://www.frederickguttmann.com*

Index

IN THE BEGINNING

Day One

As a civilization we have had an obvious interest in knowing our origin and the reason for our existence. How to go to the past to know the ins and outs of what arose when there was no human way to record what happened for the last generations?

Throughout the last two centuries, scientific advances and theories have increased considerably. Although many estimates are based on assumptions, mathematical calculations or probabilities, it is accepted that models that present the idea of the origin of the universe have, to a large extent, a high percentage of possibility of being right. However, mainstream science and much of the Abrahamic religious community supposedly do not agree on their views, and this seems to create a huge gap in ideology or popular belief. For the vast majority of Christian and Muslim groups - and even Jews - they reject the official version of classical science, arguing that the origin of the universe is already presented in the biblical book of Genesis, in its first chapter, and does not correspond to the theory which is supposedly attributed to atheist scientists. On the contrary, in Hinduism this discrepancy does not seem to exist, even the Catholic leadership, considering that the study of both currents has simply been misinterpreted, and they do not have to be incompatible.

Many refute that a book written by a Hebrew, Moses, could have had the resources or knowledge to know the origin of the cosmos and of life, but in contrast, whether supporters of creationism or panspermia (especially defenders of ufology) are – to a greater or lesser degree – convinced that a "foreign" intervention (angels, gods or extraterrestrials) provided the peoples of the past (not only the Hebrews) with knowledge that science is only now beginning to discover. If so, could the first chapter of the first book of the Hebrew Moses really deal with a fully "scientific" worldview?

בְּרֵאשִׁית בָּרָא אֱלֹהִים אֵת הַשָּׁמַיִם וְאֵת הָאָרֶץ׃

With these Hebrew characters the story of the biblical creation begins, where transcribed it would say, «*barashit bará elohim et ha.shamaim ve-et ha-aretz*», but what do all these words mean? Barashit, or Bereshit is the Hebrew name for the book of Genesis, and comes from the same root as Persian 'Bahashit' (heaven). This word means "beginning", of the form 'ba-rashit' (head), where 'Rosh' is head, beginning or leadership. The word 'Bará' means 'to create' or 'to form'. That is, ' created'. Said word is formed from the abjad letters Beit, Reish and Aleph, where the root can denote an associated hidden meaning of Bor (Beit, Vav and Reish), which is "well" (compare many other cosmogonic texts). Precisely the first word is made up of the same initial letters as the second: Bará = Bara-shit, thus being an evocation of a principle of creation or establishment (because 'Shit' would come from the voice 'Set' (establish)).

For its part, the term 'elohim' encompasses any concept related to a powerful being, a divinity or a deity, whether singular or plural. Then, the word 'Shamaim' refers to "heavens", starting from the Semitic words it is understood as a compound idea: Sham-maim (over there-waters) or Shem-Iam (name of the sea). The ancient Egyptians considered the night sky as a sea, and this was a very popular idea, so it was accepted that this celestial "sea" was full of life. While sometimes words such as 'Shamei' or 'Shamai' (sky) are used, the truth is that almost all times the plural 'Shamaim' is used, implying that beyond the celestial vault there are many "heavens". Thanks to the large number of texts that exist, we can understand that these skies are all kinds of spaces, whether physical or immaterial, and of various dimensions and universes, not limiting themselves, at all, to the common idea of many monotheistic aspects. In the case of the Sumerian version of the creation story, the origins tablet reads, "*E-nu-ma e-liš la na-bu-ú ša-ma-mu,*" referring to Nabu as heaven, and the Sa.ma.mu as the fact that it had not yet been established. Subsequently, verse 1 ends with the word 'Aretz', from which the words that refer to 'earth' come. The word Aretz derives from the Sumerian 'Eridu' (founded earth in the distance), but according to many remote sources, it seems to evoke each and every one of the physical worlds that have been produced – and are produced – in creation (the universes). The second line of the Enuma-Elish reads «*šap-li-iš am-ma-tum šu-ma la zak-rat*», referring to the "below", which in this case is called 'am.ma.tum', that is to say, Earth, had not been called by a name either, that is to say, it was not yet structured or defined; and it is important to note that the Earth is called in Sumerian 'Tiamat', but here it is not yet called that way, a connotation that reinforces the supposition that it speaks of the primordial mass from which Earth and other "earths" (worlds) were " built".

Among the fragments of an old manuscript attributed to the patriarch Abraham there are some references that already at that time implied that human existence must develop throughout eternity in the infinite worlds that populate the cosmos, and that the "angels" they appear to be evolved beings who transcended the human form to which they once belonged (that is, if angels dwell in heaven and inhabit the stars, it is possible that they would have been human inhabitants of those sidereal places): "*The worlds are infinite and man has to live in all those that exist today; but the creation continues and does not end. All the worlds communicate with each other in love and justice, and my god magnifies himself in this. All the sons of my god, whom you call angels, men were...*» (Text of the Secret Testament of Abraham). As with so much literature that already spoke of this millennia ago, man seems to be a creation that exists throughout the universe, and seems to be a consciousness (soul) that came from a previous universe and/or a spiritual universe (that ' Shamaim') from verse 1 of Genesis 1.

But the story of Moses does not begin by telling us about the original void that is so often ventilated among conventional scientific theories, but rather that the universe seems to have been constituted periodically, first from a primeval spiritual universe and, after this, the universe that we now experience. In this way, the question about what was before the Big Bang would be answered by referring to a root universe from which this cosmos comes, and perhaps many others. It is here, at this point, when Moses writes the following...

וְהָאָ֗רֶץ הָיְתָ֥ה תֹ֙הוּ֙ וָבֹ֔הוּ וְחֹ֖שֶׁךְ עַל־פְּנֵ֣י תְה֑וֹם וְר֣וּחַ אֱלֹהִ֔ים מְרַחֶ֖פֶת
עַל־פְּנֵ֥י הַמָּֽיִם׃

This next verse tells us, "*ve.ha.aretz haitah tohu va.bohu ve.choshech al-pnei tehom ve-ruach Elohim merachefet al-pnei ha.maim.*" This could be translated as, "*and the earth was in chaos*", but since at no time did he tell us how the Earth was formed, even with its resemblance to the design of the rest of the spheres that extend throughout the universe, we must ask ourselves if effectively 'Aretz' identifies our world or matter itself. In other words, Aretz can evoke the principle of atomic composition that would have been established in space, or the very scenario for future life that would develop in the cosmos. If so, what was that chaos? In theoretical paradigms, chaos is understood as the beginning of nothingness, the unpredictable, the complexity of the supposed causality in the relationship between phenomena or the reason for the fate of the universe. The ancient Greeks were already well versed in philosophies on this conception, and in all cases it seemed to identify the source or force that governs the cosmos in a certain space, under certain parameters that do not follow the model or laws of the rest of the system. In other words, the progression of the development of the universe would be accompanied by the existence of cosmic laws that push energy and order and establish the order of everything.

This means that there is a primordial force in space that seems to order all things and give them laws (gravity, electric charges, magnetism, etc.), and by means of a motor force it causes vortices that condense and repel (like a positive charge). and another negative), keeping everything by force of attraction and repulsion in a spiral. That would apply from atoms to galaxies themselves. Chaos would have been the initial part in the "vacuum" where even that energetic force or "intelligence" had not given impulse to the vibration that produced the fields, and it would still be all space in the infinite to where even that "creation" in progression would not have arrived. This version of history is similar to that of many other towns, only with different nuances, since it follows the same scheme, telling us about that "tohu va.bohu" which are also words of Sanskrit root that have a sense of "abyss" and "disorder" (absence of a principle of order or law).

It makes sense that it says 'Joshech' (darkness), if we accept that in the Big Bang theory, before this massive explosion appeared, there had to be a space where it occurred. Therefore, first there must be a "place" and then the explosion that occurs in said "place". Consequently, if the "potential" of the "material" universe was not yet there, said space was in the dark. Although other texts such as the Hebrew Jubilees, 2nd Baruch, 2nd Ezra or 2nd Enoch vary in one detail or another in the order of events, the pattern is basically the same as the one followed by Moses in Genesis 1, and at this point he already speaks of "water" and lego from "light" (other manuscripts put light first and then water). Whoever does not know the most accepted theory of the formation of matter in the universe does not know that it coincides with the Greek translation of this biblical passage. The word for 'water' in Greek is 'idor', from which comes 'hydros', and hence 'hydrogen'. Matter is made up of atoms; atoms form molecules through valences, and all this follows the law-order principle that we have been talking about. The scientific community considers that the primordial matter is the most encompassing gas in the cosmos: hydrogen.

« ἡ δὲ γῆ ἦν ἀόρατος καὶ ἀκατασκεύαστος καὶ σκότος ἐπάνω τῆς ἀβύσ σου καὶ πνεῦμα θεοῦ ἐπεφέρετο ἐπάνω τοῦ ὕδατος »

The last word of this quote, which corresponds to the Greek version, is 'Idatos' (water), in the same way 'Idor' and 'Ydros'. Water is twice as much hydrogen as oxygen. Rock crystals, water, ice and most of the gas scattered in space come, according to modern theory, from the atomic principle of hydrogen, since it is the simplest atom, that is, the primordial structure and matrix of the matter in the universe. Being so, by telling us in Genesis that there was a 'Ruach' (wind) 'Merajefet' (fluttering), it seems to indicate clearly that a spiral or vortex movement was moved or pushed in all parts of space to condense matter. Thus the charges produced the basic atom: hydrogen. Following this influence and impulse, the hydrogen would have produced condensation of water, clearly frozen due to the temperature of absolute zero. The movement would have created friction and friction energy, thus causing these vortices to condense solid matter, largely frozen water, evidently with the other sums of elements that we have studied in the Periodic Table. However, the vast majority and the base would be hydrogen, ice, water... In the midst of that darkness, what could be seen? Nothing. Without stars in space, what can you see? Nothing at all. That is what defines the Greek word 'Aóratos' in the quote above: invisible. The matter or substance, the existent, could not be seen. Jerome's Vulgate version adds that it was 'Vacua' (empty), and all these concepts coincide with the Hebrew version, and even the Aramaic, which says "*tzadia ve.rokania*" (deserted and barren). The Enuma-Elish, a Sumerian account preserved on several tablets found in the ruins of Ashurbanipal's library at Nineveh (now in the British Museum), dating back to 669 BC or 627 BC, begins by telling that «*When above, Heaven had not yet been named, and below, Earth had not been mentioned by name, nothing existed except Apsû, the ancient one, its creator, and chaos, Tiamat, of the that everything was generated. The waters were agitated in a single group and the pastures had not yet formed and the reed beds did not exist. When no star could yet be seen, none had a name when the destinations had not yet been established.*» (Lines 1 to 8). Ironically, the Earth is first a mass, and what came to be the mold or model was the 'mu.um.mu' (chaos), at that time Ti-amat, "*mu-al-li-da-at gim- ri-šu-un*» ("from which everything was generated", or "the mother of the two", referring to Apsu and Tiamat).

The phrase «*meš-šu-nu iš-te-niš i-ḫi-qu-ú-ma*» (waters in a single set) make us think, where and how were they configured? Who are Apsu and Tiamat? According to the experts, Apsu is a masculine concept alluding to the "primordial waters" and fresh water, while Tiamat is a feminine idea of the abysses and primeval chaos. This would be the equivalent of the Hebrew 'tohu va.bohu'. If we look at the etymological root of these definitions, we find that the Semitic Ab.Zu (wise father) came from Apsu, from which he later derived 'Abisu', then the Greek 'Abisos', the English 'abyss', or the Spanish 'abismo'. What's more, the southern regions or the Southern Hemisphere were called 'Abzu', as an idea of "what is below". As for Tiamat, or Tiamatu, it gave rise to the Semitic form 'Teemut', and that to the Hebrew 'Tehemoth' (abysms), whose root is precisely the particle 'Tehu' or 'Tohu', from the endiadis 'tohuvabohu' (curiously in German, this concept - defined as 'Tohuwabohu' - is a colloquialism for "chaos").

This idea of "the waters" agitated "in a single set" is in tune with other sources on the same subject of the aqueous mass that was produced from the condensation and vortex fusion of hydrogen with other basic elements that began to compose matter. And what was driving all this? That 'Ruaj' (wind, spirit). In that time, before time, is when the visible appeared, because previously the darkness closed in on everything as there were no stars, as so many sources tell us, including line 7 of the Enuma-Elish: «*e-nu-ma ilâni la šu-pu-u ma-na-ma*» (when still no star could be seen).

וַיֹּאמֶר אֱלֹהִים יְהִי אוֹר וַיְהִי־אוֹר׃

Here Genesis 1 tells us, "*ve.iamer Elohim iehi aor va-iehi-aor*" referring to the fact that it was light, and it was light. How was that light produced? In line 9 of the Enuma-Elish this is also mentioned: «*íb-ba-nu-ú-ma ilâni ki-rib ša-ma-mi*» (Then, the stars were made visible in the middle of the sky). Although it is usually believed in creationism that the universe was made in 6 periods (for some this is even literal, or in the best of cases a reference to 6,000 years), but the first period did not begin to be marked but rather from this time, so the above happened before this definite time. In this order of things, only in the 4th period would the stars have appeared, not in the first. However, if the stars were supposed to appear in the 4th period, what was that "light" (Aor) that took place before the time clock began to mark?

Moses did not contradict himself, and in another work of his authorship, recovered from the Qumran caves, he tells us that the eternal, «*For the first day that created the heavens and [what] is above the earth and the waters and all the spirits that serve before him [...] and the abysses of darkness, [...] (and night), and light, dawn and day, [...] seven great works that He created on the first day*." First day? The Hebrew term that is translated as 'day' is 'Yom', but it identifies a period and location where light dominates. The word Yom is associated with the Greek "aeon", as Age or Era. The Castilian word "day" derived from the Greek, this being a qualifier for the name Zeus (God), and this reminds us that the ancients related the gods and angels to the stars, just as others translate line 9 of the Enuma Elish: «*Then the Gods were created in the middle of the sky.*»

In the version revealed to the Hebrew scribe Ezra, «*then there was the spirit, and darkness and silence were everywhere, the sound of the man's voice was not formed. Then you sent a light just out of your treasures, so that your work could appear*." (2nd Ezra 6:39-40)

The Big Bang? It seems a description that both supports the classical version of the origin of the universe and that of other cultures, which maintain that everything arose from "another place" and was provided from there by means of a great light with its spiritual or invisible forces. who started the creation. The Chinese and Hindu versions of this story are similar, considering that things came out of a kind of "primordial place" that opened up and produced everything: the Vedas describe it as an egg that cracked and from there came the "deities". which produced heaven and earth. It is admissible to consider that when saying "earth" it does not refer explicitly to this planet at first but to all those in the same conditions, which are not asteroids or comets or stars, and meet certain parameters that include them in the category of "planets". In fact, in the version of the prophet Enoch, the Earth was produced from a thick dark liquid mass that was fragmented into 8 parts, leaving the one containing the densest waters configured as our planet Earth, while the other 7 became the other planets of the solar system.

וַיַּרְא אֱלֹהִים אֶת־הָאוֹר כִּי־טוֹב וַיַּבְדֵּל אֱלֹהִים בֵּין הָאוֹר וּבֵין הַחֹשֶׁךְ

Verse 4 tells us, «*ve.irá Elohim et-ha.or ki-tob ve.ibdel Elohim bein ha.or ve.bein ha.joshej*» , referring to the fact that Elohim sees the good of this "light" and performs a distinction or separation between light and darkness. What at first had been coming together was now separating; the substances obeyed the force of gravity and the centrifugal force. According to the words of the prophet Enoch, the god of the Hebrews took him out of the Earth and showed him how he made the universe, and narrating about this apparent "Big Bang", he tells him: «*I ordered [...] that visible things should come down from the invisible ones, and Adoil (light of creation) came down very grand, and I contemplated him, and he came down. He had a great light belly. And I told him: Get rid of it, Adoil, and let the visible come out of you. And he came to undo, and a great light came out. And I was in the midst of the great light, and just as there is born light from light, a great age came forth, and showed all creation, which I had thought to create. And I saw that this was good. And I set up for myself a throne, and took my seat on it, and said to the light: Go therefore up on high, and set thyself high upon the throne, and be a foundation for lofty things."* (2nd Enoch 25:1-5)

Enoch seems to describe the story from another perspective, arguing that first it was this light that emanated from who knows where and broke up or exploded making the invisible appear as visible, but after this God commanded that solid matter emerge: «*And I summoned the very I came down a second time, and said: Let Arjas (spirit of creation) come out hard, and he emerged hard from the invisible. And Arjas came out, hard, heavy, and very red. And I said: Open up, Arjas, and let it be born of you there, and he fell apart, an age came forth, very great and very dark, bearing the creation of all low things, and I saw that it was good. ...»* (Chap. 26:1-3)

Enoch later speaks of this 'separation', saying that God told him: «*And I ordained that there should be taken from light and darkness, and I said: Let it be dense, and it was thus, and I parted it out with light, and it became water, and I parted it out with darkness, below the light, and then and made the waters firm, that is, the bottomless (bottomless), and I I made a foundation of light around the waters, and I created 7 circles from within, and imagined the waters as wet and dry crystal, which is to say as glass, and the circumspection of the waters and the other elements, and I showed each one of them. they their way, and the 7 stars each one of them in their Heaven, that they go like this, and I saw that this was good. And I separated between the light and the darkness, which is to say in the middle of the water here and there, and I told the light that it must be day, and the darkness that it must be night, and it was late, and tomorrow was the First Day."* (Ch. 27)

Remarkably it is impossible for conventional history to think that someone gave such descriptions more than 4,500 years ago, so this prophet could not be making it up. His description, archaic - it is understandable - due to his time and limits of his own language, are brief enough to understand that he tells us how after the appearance of that light that breaks off throughout the universe, or the Big Bang, matter it condenses in the middle of space and appears as abysmal masses of condensation based on hydrogen. His description of how light and dark separate to shape water as such can only be explained as the structuring of water due to the atomic valence principle of the gases that make up water and also produce electricity. Since these words did not exist in remote antiquity, it was to be expected that they simply referred to this as the invisible that sustains things or is the essence itself, that is, the Ruaj (wind, spirit). That Ruaj that "moved" on the water being the law or force that was "curdling" the matter.

In addition to this, the mention that Henoc gives in chapter 27 of his Second Book regarding water and its shape suggests that he was told that the invisible and motor force caused the masses to consolidate into "circles" by rotation. As you will see if you review the beginning of this Article, you will see that I already commented that the word 'Aretz' seemed to designate the physical worlds created, at least, in this universe. Although Moses only spoke about it quickly, Enoch qualifies it in a formidable way, letting us know that God already thousands and thousands of years ago had told him how the motor force that comes from Him produced the Big Bang, the forces of the universe, the charges that created the atoms and later the molecules and, consequently, the gases from which the water (liquid or not, pure or not) was produced.

Although at first glance it seems that Enoch describes the formation of what could be the planets of our solar system, it is in chapter 30 when he tells us about these 7 space bodies ordered to fix themselves in their respective orbits, and what can be interpreted is that the 7 designate a pattern. When he says that he created 7 spheres one inside the other, one must assume that he is talking about a worldview of 7 heavens or dimensions, and/or a pattern of 7 levels subject to the geometry of the sphere as an elemental model of the universe. This same model would have given rise to the structure of our solar system, although, according to this description, and those that appear to be provided by the Sumerian sources, it would have been different from how it came to be later, having been reorganized thousands of years later to have the distribution that today they own Moreover, when talking about these 7 spheres and each one in its own sky, it suggests that there are at least 7 heavens or that there are 7 existential realities in the cosmos, but above all it reminds us of the 7 planets of the solar system in their orbits. Does this mean that you would only talk about the creation of our system? According to these words, God would call this the "heavenly circle", and he affirms that he ordered that in those "spheres" or "circles" there would be "luminaries", and following this pattern he made everything in "all the heavens". This means that the scheme that the Creator used was that of a fractal based on a system of 7 at all levels.

וַיִּקְרָא אֱלֹהִים ׀ לָאוֹר יוֹם וְלַחֹשֶׁךְ קָרָא לָיְלָה וַיְהִי־עֶרֶב וַיְהִי־בֹקֶר
יוֹם אֶחָד

Despite the fact that many of these explanations have already been dealt with in the work 'The Rebellion of Sakla I, The Abyss', there is a lot of information that it is always good to review, restructure, update and expand. Here verse 5 tells us, «*ve.ikra Elohim la.or Yom ve.la.choshech kara lailah ve.iehi-ereb ve.iehi-boker yom achad.*» We could translate this as that the designation or identity of the 'Aor' (light) then becomes 'Yom' (day, aeon, era), while that of 'Joshej' (darkness) becomes 'Lailah' (evening). At this point, the whole set is known as 'Yom Ajad' ('Day One' or 'Eon Ajad'), and if we remember pattern 7, it is reflected here again, since 6 is spoken of as formation time, and 7 as the end of this cycle to follow others.

Day Two

«*Before the beginning there was only one consciousness, that of the Eternal whose nature cannot be expressed in words. It was the only Spirit, The Generating Being that cannot be reduced. The Unknown, Unknowable A melancholic loner in deep pregnant silence. The name that was pronounced cannot comprehend this Great Being that, remaining nameless, is the beginning and the end, beyond time, beyond the reach of mortals, and that in our simplicity they call God. He who preceded all existed only in His strange room of uncreated light, which remains ever unquenchable, and can never behold it by any comprehensible eye. The pulsating drafts of light, the eternal life in its keeping had not yet been released.*>> (The Book of Creation, The Kolbrin. The records of Celtic culture).

This majestic explanation of the Kolbrin wisely summarizes everything previously said, and broadens the vision of this story: «*He knew alone, he was without contrast, unable to manifest himself in nothing, since everything within his being was, [he] Unexpressed potential. The greater circles of Eternity had not yet come out, to be thrown out like the centuries of the centuries of the existence of [the] substances. They were to begin with God and return to Him completing in infinite variety and expression. Earth did not yet exist, there was no wind with the sky above them; high mountains were not raised, nor was the great river in its place. Everything was formless, motionless, calm, silent, empty, and dark. Nameless had been named and there were no foreshadowing destinies.*"

The old Greek accounts tell us that «*Before there was the great sea and the fertile land, and the blue sky that covers the world; before nature –which our eyes see and all our senses help to accept- lived as it does now: organized, plastic, wise, powerful; Before all that, it was Chaos: the rough, shapeless mass that made up the universe. In the beginning, what existed was inert, says Ovid, a Latin poet. He was a dead weight. Lots of disparate items. At that time, no light gave the world warmth and clarity. Neither the Sun nor the Moon traveled the celestial vault, transforming each day into a new day, and each night into a clear night. The Earth was not yet suspended in the air, balanced by its own weight...*

...And Amphitrite, the queen of the sea, had not yet extended her sweet arms to the shores. Earth and Sea were an indistinct mixture of life and agitation. The ground had no density. The sea did not flow. The air had no light. Nothing had its own shape. And inside that unique mass, the constant battle of the opposite principles was engaged: the cold fought the heat; humidity against drought; lightness against weight. Little by little, an intelligent seed, an ordering god, emerged from Chaos. He defined (delimited) and harmonized (balanced) everything, according to his sovereign will. Peace was made in the universe. But the spark of conflict remained lit forever, because order, limits and balance are not static...»

The Quiché Indians also tell us the Mesoamerican version of this story: «*Here is the story of how everything was in suspense, everything calm, everything motionless, everything peaceful, everything silent, everything empty, in heaven, on earth. Here is the first story, the first description. There was not a single man, not a single animal, bird, fish, crab, wood, stone, cave, ravine, grass, jungle. Only heaven existed. The face of the earth did not appear; Only the limited sea existed, all the space of the sky. There was nothing gathered together. Everything was invisible, everything was still in the sky. There was nothing built. Only the limited water, only the calm sea, alone, limited. Nothing existed. Only immobility, silence, in the darkness, in the night. Only the Builders, the Shapers, the Dominators, the Mighty Ones of Heaven, the Procreators, the Begetters, were on the water, scattered light. [His symbols] were wrapped in the feathers, the green ones; their [graphic] names were, then, Feathered Serpents. They are great Sages. Such is heaven, [such] are also the Spirits of Heaven; Such are, tell yourself, the names of the gods. Then the Word came; He came here from the Dominators, from the Mighty Ones of Heaven, in the darkness, in the night*

[...] they united their words, their wisdom. Then they showed themselves, meditated, at the moment of dawn...» (Popol Vuh. Chapter 2)
In their time, the Nordics already said that the world began as an icy mass called Niflheim, and with it came the world that we can define as that of magmatic activity, which they called Muspelheim, as well as the chaos or abyss that existed between the two: Ginnungagap. We also have the idea of the universe being produced with its God from a golden or brilliant womb, a cosmic egg called Hiranyagarbha, which is narrated in the sacred Bhagavata Purana of India. This is how the heavens, worlds, their orbits and dimensions, planes of reality and even spaces and parallel universes would have been produced, and in them the great God, in many forms and representations, who inhabits everything, "*the one who dresses (using as mantle) the solid stones of Heaven. And He also chose as many as please Him»* (Yasna XXX, the Avesta of Zoroaster. Ancient Persia) The manuscripts of the Egyptian library of Nag Hammadi tell us about these stories in their various treatises, exposing that the "gods" emanated in chaos, in that darkness and watery mass. Although, in Coptic and Koine Greek the word for 'gods' was more precisely 'exoisias' (authorities) and 'arjais' (archons, principalities), which later created the 'dynameis' (powers, forces, energies) and with who represent those powers of Fate. Ironically, although the Hebrew word 'elohim' is translated as "gods" or "God", it really means "mighty ones", being able to evoke "powers" or "forces" behind the movement of the material world.
In another treatise of these Egyptian Christian records, we find: <<*Sem, since you are of unmixed power and you are the first being on earth, hear and understand what I am about to tell you, [and] for the first time about the great powers that were in existence in the beginning, before the present. There was no light or darkness and there was no Spirit between them. Given its root they fell into oblivion - the one that was the unbegotten Spirit - of which I reveal the truth to you, about the powers. The Light believed itself to be complete from hearing and speaking. They were united in one way. And the darkness was the ruaj (wind, spirit) in the waters. His mind*

was wrapped in a chaotic fire. And the Spirit between them was a soft and humble light. These are the three roots. They each reigned in themselves, alone. And they covered each other, each with his power.>> (Paraphrase of Sem 1:19 to 2:10) All of this agrees with the rest of the versions that deal with this story of the origin of the universe, since the idea of an Eternal and Immeasurable God who existed before the universe and from whom everything comes, it is present in many cultures.

«*Death was not then, nor was anything immortal. That without breath, breathed by its nature: apart from it, whatever it was, there was nothing. Everything that existed was empty and formless. What was above or below then? Who certainly knows and can declare it? Where was this creation born and where does it come from? The later gods were the production of the world. Who knows then where the existence began? He, the first origin of creation, whether he formed it or not, whose eye controls the world from the empyrean, He certainly knows, or perhaps not...*» (Rig Veda, records of India) These peoples called the gods "gods". powers and stars that emerged with the "Big Bang".

וַיֹּאמֶר אֱלֹהִים יְהִי רָקִיעַ בְּתוֹךְ הַמָּיִם וִיהִי מַבְדִּיל בֵּין מַיִם לָמָיִם׃

The story of Moses, in its verse 6, tells us, <<*ve.iamer Elohim iehi rakia ba.toj ha.maim ve.iehi mabdil bein maim le.maim*>> , that is, that there was a space or void between the water, understand that already existing there. He adds that this is done by separating between "waters to waters", that is, waters over waters. The para-biblical Hebrew text of 2nd Esdras (chapter 6:41) tells us: «*On the second day you made the spirit of the firmament, and you ordered it to be divided in two, and then to make a division between the waters, which could rise on one side, and the other will remain at the bottom.*» This description is clearly an allusion to the formation of the atmosphere, at least in our world and those that are similar spheres. How would this occur? It is a logical action that responds to the law of gravity: some elements are heavier than others, and the lighter ones (gases) overlap the heavier ones (liquids), and these even heavier ones (solids). Jubilees 2:4 confirms this fact, arguing, from the words of Moses: «*And on the second day he created the Rakia in the midst of the waters, and the waters were divided on that day; half of them passed over and the [other] half of them went down below the Rakia (which was) in the middle over the entire face of the earth. And this was the only work [that] (God) [had] created on the second day*."

The thousand-year-old Emerald Tablets of Thoth speak to us of "*the Lords of Amenti, they lords of the Children of the Morning, Suns of the cycles, Masters of Wisdom*." (Table III) Dyehuty (Thoth), the Egyptian "god" of writing, claimed that these lords of Amenti controlled chaos so that the universe would not be dissolved, and, according to him, the existence of chaos was necessary to maintain a balance in the cosmos. Dyehuty referred that these immortals of Amenti (the halls of the path to the afterlife) were lords over the "children of the morning" and in turn "suns of the cycles". The Idumean sage Job wrote, some time after Abraham, "*when all the morning stars sang together and all the sons of God rejoiced...*" (Job 38:7)

Did the stars sing praise? In Hebrew it says 'Kokabei Boker' (morning stars), the same ones that Dyehuty seems to mention, and of whom he added, «*Many of the stars I passed on my journey, many of the races of men in their worlds; some rising high as the morning stars, some falling low in the blackness of night. And he also wrote that "Many ages ago, the Morning suns descending, found the world filled with night, there in that past, the fight began, the ancient Battle of Darkness and Light.*» (Table VI)

Now, the Hebrew word Yom (day) alludes to an era and a god, as encompasses the Greek word 'Aionos' (aeon, era, century, age), also 'Laila' (night) has its own deep meaning. Its Sanskrit root, 'Lila', identifies it as the concept related to the eternal struggle in the cosmos, since darkness emerged. For the Vedas, the Lila was the eternal cosmic battle between good and evil, the struggle between order and chaos, light and darkness, justice and injustice. The very name of the demoness Lilit, from the Hebrew legends, takes its name from the same root, being the idea of evil, darkness and the demonic. Another important point here is that the narrative of Genesis 1:5 tells us about a 'Boker' (morning) and an 'Ereb' (evening) that configure that first eon or initial era, but, if the sun is not supposed to was, what were they that morning and that afternoon? Furthermore, it does not say "night", but that the day would only be according to the period of light, the principle of 6 hours in the morning and 6 hours in the afternoon? Consequently, he does not speak of a 24-hour day, but of a light cycle. If you were talking about a consecutive period of actual days, you would have included the night. This agrees with the other sources that expose that this light was not only cosmic (from spheres of light), but from beings of light that govern the positive side of that Ying-Yang, or "Lilá", as opposed to the chaotic ones that govern the dark side.

We can conclude here, without fear of being wrong, that these "stars" were extraterrestrial beings from another dimension, angels from higher spheres of reality who fight against darkness and its forces, and those "suns" or "sons of god" are equally a high degree of supernatural intelligence that operates on order in the cosmos. In the "dawn" periods of civilization they were here, some of them having been deserters just before the flood. Genesis tells us that the Yom is divided into two sections: Boker and Ereb. Those of Boker are of maximum light and of that era, but, what is that of 'Ereb'? The twilight of that era (cycles that began and ended with decisive episodes), of which the ancient Greeks recorded when they said that in the remote era of the world, 'Erebus' was thrown into hell, where he became the horde of demonic spirits that from that remote past devastated the world.

Enoch's version is even clearer about this story, and tells us: <<*And then I made the celestial circle firm, and caused the low waters which are under Heaven to collect themselves together, into a hole, and that chaos became dry, and it became so. Out of the waves (wave) I created hard and great rock, and from the rock I filled the dry, and the dry I called earth, and the middle of the earth I called the abyss, which is to say the bottomless (unfathomable), I collected the sea in a place and limited together with a yoke. And I said to the sea: Behold, I give you your eternal limits, and you must not break yourself, let go of your component parts. So I quickly made the firmament. And the evening and the morning were the Second Day.>>* (2nd Enoch 28:1-4) The so-called "celestial circle" that was formed when the parts of lighter water (gases) rose, from the rest that remained below, was that void (Rakia) between both, and the rest of matter from the bottom covered what was originally a hole.

This is what is also told in the Enuma-Elish, where the story is described as a pitched battle between gods to refer to the way in which the solar system was formed. According to the long description given by the Sumerian tablets, various impacts of space bodies and formation processes led to the establishment of the solar system as we know it, but the primordial mass Lahmu and Lahamu changed state and "spirits" emerged and subjugated. Precisely the words Lahmu and Lahamu give rise to the Hebrew word Lejem (bread), due to the idea of a pasty mass that is obviously later heated to obtain the precious food. The earth as a quagmire and also burning (still hot and with many volcanoes and lava) began to produce its atmosphere in a long cycle of changes with the consequent change processes, pressures and revolutions of the terrestrial sphere. That idea that Enoch describes, of chaos becoming dry or solid, agrees with the other passages in the story of Moses, so we can argue that the changes happened one after the other, but the sequence was not necessarily according to the end of an era for start the next one, but even many starting before the other one is finished. How does this whole story differ from the scientific version about the geological formation of the Earth? Actually practically nothing, except that these peoples already knew it thousands of years before current science.

Biblical translators translated 'Rakia' in various ways, from 'vault' (Reina Valera 1995) or 'expansion' (Reina Valera 1960), but since the translation of the Seventy the idea of 'Steréoma' (firmament, firmness, firm), as Jerome did much later to the Latin: 'Firmamentum' (Vulgate version). There are many texts, whether Hebrew or not - especially from the prophet Enoch - that address the issue of how the "spirit" of God formed the atmosphere - or "heaven" according to "biblical" vocabulary - and designed it in this perfect spherical shape covering the surface of our world. The word itself, Rakia, derives from the root 'Rek' (void), since in a way the layers of the atmosphere, although filled with gas, appear to be a vacuum above us, as if we were inside a large transparent balloon. filled with air or helium, because the gas is invisible, even though it takes up space. Now, all that gaseous mass did not cease to be part of those primordial waters, even though at first glance our perception of things does not make us imagine this as such.

This situation even responds to the hydrological cycle, where the magical clouds play a transcendental role, and that invisible "spirit" that motivates life carries the wind currents from the south and north, east and west, whether in the atmosphere or in the ocean floors; it maintains the clouds that were born with the atmospheric formation at their height thanks to the pressure and the climate, it promotes the precipitations that nourish the mainland, it carries the fresh water in descent to the sea and again the sublimation in the ocean brings the particles back liquid towards the clouds. With all reason the text says that "*the waters are separated from the waters*".

וַיַּעַשׂ אֱלֹהִים אֶת־הָרָקִיעַ וַיַּבְדֵּל בֵּין הַמַּיִם אֲשֶׁר מִתַּחַת לָרָקִיעַ וּבֵין
הַמַּיִם אֲשֶׁר מֵעַל לָרָקִיעַ וַיְהִי־כֵן׃

In the seventh verse of Genesis 1, the Hebrew text tells us, «*ve.ias Elohim et-ha.rakia ve.ibdel bein ha.maim asher mi- tachat le.rakia ve.bein ha.maim asher meal la.rakia ve .iehi-ken*", which in a few words means that this "firmament" was made and the "waters" that were under the firmament were separated, which remained that way, but also the "waters" above, on that " empty". It can be assumed that certain "waters" ended up beyond this world, in the sky, as some of us have once suggested, giving formation to other planets (as is the case of the "aqueous" or mega "gaseous" ones that are beyond the Earth: Jupiter,

Saturn, Uranus and Neptune (planet that, however, is identified as "the sea of heaven" in the Vedic culture, and that the Greeks called 'Poseidon')). However, what can be considered the easiest is that we are talking about O3 (the ozone layer). Just that cape? No, the context of the atmosphere as a gaseous mass.

So, just as before Day One (first era defined, Aeon One) was the spiritual universe, later the "void" in this universe, then the appearance of light (Big Bang, stars, etc.) and then the matter being configured, in the Second Aeon the "Aretz" or physical masses were structured, which thanks to the law of gravity took the spherical shape and separated its substance by weight, giving rise to the outer planetary layers: atmosphere. Not all planets seem to follow the same pattern and be a paradise, but the idea suggests that they follow the same pattern (which according to the laws of physics would make a lot of sense).

וַיִּקְרָא אֱלֹהִים לָרָקִיעַ שָׁמָיִם וַיְהִי־עֶרֶב וַיְהִי־בֹקֶר יוֹם שֵׁנִי׃

Verse 8 tells us, «*ve.ikra Elohim la.rakia shamaim ve.iehi-ereb ve.iehi-boker iom sheni*», which defines that the 'Rakia' received the title, name, designation or identity of 'Shamaim'. This is how this Second Aeon ended, but the problem here is that before the "days" began, it had already been said that the Shamaim, as well as the Aretz, had been created. What's more, explanations were given about the state of that matter, or 'Aretz', which came after the Shamaim, so this “other” Shamaim seems to emulate the first one or be an “inferior” image of it. If you observed in the previous quotes, the "name" that was before all things and from which they emanate was constantly spoken. Precisely "name" in Semitic languages is 'Shem', which is also read as 'Sham' (over there), and which gives rise to the word 'Shamaim'. Although in Hebrew "names" is 'Shemot', in the feminine plural, not in the masculine plural, it should be noted that 'Shamaim', or 'Shemim' was an idea of 'names' (identities, destinies) projected from that original Shem, being Said original Shem the engine of all the others.

Many manuscripts tell us that the sky is full of skies, and what do those skies consist of? Obviously this is a precarious definition that encompasses what we now know how to define the universe with many words. If this Shamaim is all the space that encompasses the atmosphere of the planets, their orbits, their dimensions and/or all of the above, these descriptions make a fascinating sense, since they fit perfectly with all the scriptures that address this matter. Even the descriptions of "vortices" or "spirals" explain that movement of the "spirit" flowing in everything and motorizing everything: «*As you see the power of the whirlwind collecting the dust of the earth, and driving together, know that in the same way the ji'ay, a'ji and nebulae are gathered in the firmament of the heavens; by the power of the whirlwind I can create the corporeal suns, moons and stars. And I told the man to name the whirlwinds in the Etherean firmament, and he named them according to their shape, which he qualifies as vortices and wark.*" (Oahspe, revelation of 1882)

Day Three

When the atmosphere was formed, the mainland also began to appear. Enoch's version told us that at the end of the Second Period, just as the Earth's gas cover was formed, the rocky mass that would build Pangea emerged from the depths. While this was happening, according to the version of this prophet, angelic beings were created, and one of them rose up and dominated the abysses: «*And for all the armies of heaven I imagined the image and essence of fire, and my eye looked towards the very hard, firm rock, and from the flash of my eye the lightning received its marvelous nature, which are both fire in water and water in fire, and one does not put out the other, nor does the one dry the other, therefore both lightning is brighter than the sun, smoother than water, and firmer than hard rock...*

... And from the rock I cut a mighty fire, and from the fire I created the orders of the armies of tens of thousands of non-carnal angels, and their weapons are fire and their armor a burning flame, and I commanded that each one should stand in his order. And one of the chief custodians of the angels, obstinate (twisted) with custody downwards and promoted [an] impossible plan: to erect his throne above the Earth considering his strength compared to mine, and the consequences of his puffiness with his angels, and he was [and] prosperous unfortunately on the face of the Abyss, always." (2nd Enoch 29:1-4)

וַיֹּאמֶר אֱלֹהִים יִקָּווּ הַמַּיִם מִתַּחַת הַשָּׁמַיִם אֶל־מָקוֹם אֶחָד וְתֵרָאֶה
הַיַּבָּשָׁה וַיְהִי־כֵן

The version of Moses in Barashit (Genesis) 1:9 says: "*ve.iamer Elohim ivakú ha.maim mitajat ha.shamaim al-makom ajad ve.teraeh ha.ibashah ve.iehi-ken"* , which refers to gather the aquatic or liquid mass that was under the sky (that is, what was under the atmosphere) in a single place, and the dryness was visible. In other words, the most solid part was consolidated on dry land, but how, if the water covered everything? In 2 Esdras 6:42 he tells us that "on the third day you commanded that the waters be gathered in the seventh part of the earth: six parts you have dried up, and you kept them, with the intention that some of these [were] carved [and] were planted of God and could serve you." The objective, according to the text, was that that 7th part of the Earth would be solid land to become cultivated, and consequently, for the purpose of serving humanity. But, if humanity did not yet exist, at least on this globe, it is necessary to assume that humanity was destined to come to this planet, at least starting from this point, or from this moment. The idea that the earth makes up only a 7th part of the earth's mass would not have been accepted by science if it were not for the discovery of the subterranean oceans that was made a few years ago. Thanks to this finding, it was learned that this planet is not strictly a mass of rock covered with water, but that this distribution is not precisely uniform.

Jubilees 2:5-7 - a text also attributed to Moses and found in the Qumran caves - states: "*And on the third day he commanded the waters to pass from the face of the entire earth in one place, and the dry land appear. And the waters did as he commanded them, and receded from the face of the earth in a place outside this Rakia, and dry land appeared.*" The question of how this would have occurred naturally is part of the enigma of plate tectonics. Although it is really a theory, what various scientific proposals suggest is that the force of gravity and the change in pressure and temperature of the elements of the lithosphere with respect to the asthenosphere cause geological changes and continental displacement. If this position is correct, it coincides with the approaches previously commented, since the vortex forces, the force of gravity and the changes in temperature and pressure raise the light parts and sink the heavy ones in a permanent cycle. This is how mountains and continents have been created, and the first of them: Pangea.

While the rock was pushed up by these forces, the water moved to the sides, in the direction of the space that touched it (because two things cannot occupy the same space at the same time), giving "birth" to the ocean, maritime mass the only one that has received the name of Panthalassa. According to the accepted chronology, all this was the end of the Paleozoic period, but the end of Yom Shlishi (Third Day) had not yet arrived, but, beginning the Mesozoic era, life "emerged": «*And thus I made all the heavens. ; and it was the Third Day", says Enoch, referring to all his previous string and to the creation of the beings of those worlds and their dimensions, and adds that "on the Third Day I ordered so that the Adamah would abound with large and prosperous trees and the mountains, all of sweet grass and all seed (seed) on the scattered; and I put a garden and closed it, and guarded [with] nurturing-attentive angels of fire.*» (2 Enoch 30:1) Jubilees also speaks of this, adding that "*that day has created for them all the seas according to their different meeting places, and all the rivers, and the gatherings of the waters in the mountains and on all the earth, and all the lakes, and all the dew of the earth, and the seed that is sown, germination and all things, and fruit trees, and the trees of wood, and the garden of Eden, in Eden and everyone. These four great works that God created on the third day.*" (Ch. 2:7-8)

Both writings tell us that when the mainland emerged, plant life appeared in its entirety - or so it is implied -, including the Garden of Eden, but in a progressive period that could last for a long time. But if the earth's surface was still united, and the continents were beginning to separate, where was that garden? to the east? Taking studies on Pangea as a reference, the "east" was the Tethys Sea (or 'Tethys'), unless it spoke of Eurasia to the north or India and Australia to the south. This suggests whether it is perhaps coinciding with the Oahspe, which affirms that the first great continent - called there 'Waga' - was dissolved, and from this specific portion of land where this particular garden was located, what we now call 'Japan' would remain. '. But there is another enigma to discover, where did the seeds of the trees come from? We know that there are plants and trees that strictly depend on human care, and also that the seeds come from somewhere. It is the same story of, "what came first, the chicken or the egg?" Logically it must have been the chicken, but in this case, where did the trees, shrubs, plants and other herbs come from? This is where the theory of Panspermia comes in, first formulated by the Swede Svante August Arrhenius, a noble prize in chemistry, and defended by Hermann Richter (although it was the astronomer Sir Fred Hoyle who made it popular).

According to the theory of natural Panspermia, life came to this world from space, and could come from one or more asteroids that had spores or frozen cells that, when they came into contact with the ground and favorable conditions on our globe, produced the forms subsequently existing. This would be like a less conventional theory of Evolution, except for the problem that, as Bio-Genesis points out, life is produced in a similar way that brings it into existence, so even if life had come from " outside" that does not explain the variety of species, intelligence or their level of development, unless a Noah's ark of plants and animals had fallen from the sky. In addition to this, life, although "life is sought", to "survive" - worth the redundancies -, depends on adequate conditions to be produced, and theoretically bacteria would not survive the extremely high temperatures and the forces involved in an impact against Earth. It is believed that extremophile bacteria would overcome this barrier, and by recreating this scenario with organic molecules, such as amino acids, it has been found that not only are they not destroyed, but they begin to form peptides.

The most coherent position is the one defended even by Ancient Astronaut theorists, who, supported by ancient testimonies and records, maintain that life was deliberately brought to our world from other confines, especially from the Orion systems. This is how the theory of Directed Panspermia makes great sense and can satisfactorily explain the dilemma of the origin of the cell and, perhaps, that of complex organisms. 2 Ezra 6:43-44 adds: "*For as soon as your word spread the work was done. Because immediately large and innumerable fruits were produced, and many sweet and pleasant to the palate, and flowers of unalterable color and wonderful-smelling odors, and this was done on the third day.*" The Sumerian version of this story also coincides, implying that the "gods" or beings of light of the universe saw that this orb was already ready to support a paradisiacal garden for the gods. The relationship found when looking at the Mesopotamian definitions for these concepts is curious, such as the contrast between "steppe" ('Edin' in Sumerian; 'Serum' in Akkadian) and "irrigated land" ('Gan' in Sumerian; 'Eqlum' in Akkadian). Although, in Hebrew, 'Gan Eden' means 'garden of Eden'. Also 'Edin' means "plain" or "flat place beyond cultivated land". It is curious that the Indo-European word with the same consonants, 'GHeN' or simply 'Ghn', probably derives from the same stem as the Sumerian form Gan (irrigated land), or Hebrew (garden, orchard). The Sanskrit form Ghen precisely means "chaos", which later became the idea of "disorder", but came to Greek as 'Gen' (Earth), and its variants: Gi, Ge, Gin, Gea, Gis, Gei, Gii. For its part, the word Eden passed into Hebrew as a colloquialism for "paradise", which in turn derives from the Greek 'Paradeisos', and this from the Persian 'Pardes'. It is even conceivable that the root itself had the same origin that gave rise to the Hebrew form Gai, or Gia, meaning "valley." All this leads us to the logical conclusion that our sphere was a watery world that stood out as a paradise when the mountains and mainland emerged, and exuberant plant life soon filled everything, and it seems especially to the east.

וַיִּקְרָא אֱלֹהִים ׀ לַיַּבָּשָׁה אֶרֶץ וּלְמִקְוֵה הַמַּ יִם קָרָא יַמִּים וַיַּרְא אֱלֹהִים
כִּי־טוֹב

In this verse 10 the text says, «*ve.ikra Elohim la.iabeshah Aretz ve.la.mikuh ha.maim kará iamim ve.irá Elohim ki-tob*», referring to the fact that the dry expanse of the world received the name 'Aretz ' (Earth, 'Eridu' in Sumerian), while the area where all the waters were gathered in a single mass, was called 'Iamim' (Seas). According to Sumerian history, Eridu was the first city that was established on Earth, the place that was initially founded when the surface was found to be stable and habitable. The book 'Mesopotamia and the Ancient Middle East' (1992) evokes the words related to an ancient Sumerian inscription, which says: «*Not a reed had grown; a tree had not been created; a house had not been made; a city had not been made; and the lands were sea, when Eridu was created*." Several times there has been this description that the world was mostly water, and the first part that stood out was this plain, which seemed to stand out from the area of Mesopotamia to Japan; later the rest of Pangea was coming out "afloat".

The legends of the Akpallu or Anedoti, of the god Oannes and other Babylonian narratives also speak of the aquatic world, of the era of the reign of the sea god Ea (later ruler on dry land as Enki), who built his first settlement offshore, analogy from the Greek story of how Poseidon built Atlantis. In fact, Enki would have been responsible for the creation of the Mesopotamian rivers, through a long process of drainage, water diversion and canals, similar to the Egyptian myth of Ptah, creator of the Nile (some believe that Enki, Poseidon and Ptah were the same person). But if now the Aretz (mainland) and the Yamim (the total oceanic extension) were clearly defined, with those names, how is it that the beginning of Genesis comes to say that it was then that the "Aretz" was created? and not three "aeons" later? There were some Shamaim and an Aretz that existed – and probably still exist – before the existence of this universe, and the Heavens and the Earth that now exist are images of those that first existed before time and space. In what proportion could this happen? How many times could this have happened or will this pattern be repeating itself in this or more universes? An eternal heaven and earth create a heaven (dimensions, planes, orbits) and an earth (matter) in the universe and these heavens and earth create many like ours: planets (Aretz with their Shamaim).

The oldest Sumerian tablet found delights us with these words: «The reptiles truly descended. The earth is resplendent like a well-watered garden. At that time Enki and Eridu did not appear. Daylight did not shine, moonlight did not emerge. The planets weren't even there yet, but the Earth had already begun to blossom and it was a beautiful, beautiful garden. Without sunlight, how did plants grow? How did they photosynthesise? Without light, how was this process of changing light energy into chemical energy to produce carbon carried out? By geothermal energy absorption? By synthesis of inorganic matter? By radiation derived from cosmic rays? None of this is entirely clear, but there are many records of plants and trees growing in conditions of almost, or at least to a great extent, permanent absence of light. They could grow and go into dormancy, but for the mere fact of growth to occur they need to follow a light source. The Sumerian text holds that the land was "resplendent," but seems to use this epithet in the sense of describing the beauty of the plant habitat.

One hypothesis, although somewhat daring, would be to consider that the possible sowers of plant life had brought seeds capable of absorbing the energy of geothermal energy or cosmic radiation, or had introduced seeds created with this capacity. All of this sounds a bit crazy, but one factor that stands out from plants and trees without light is that their growth increases more than normal due to a normal reaction whose purpose is the need to extend their trunk in search of light. If something has been instilled in us since we were children at school, it is that the Mesozoic era stood out for the outbreak of life forms of a supernatural size, whether in plants and trees or in animals.

The balance of carbon on our planet comes mainly from photosynthesis carried out in the aquatic environment by algae, cyanobacteria, red bacteria and purple bacteria and green sulfur bacteria. This could explain why the conditions were propitious since before the appearance of terrestrial plants, even when there was no sunlight. The thousand-year-old Egyptian text of Hermes Trismegistus, 'Corpus

Hermeticum', tells us that "*there is a place, beyond Heaven, a place without stars and separated from all corporeal things. There is another Dispenser that is between Heaven and Earth, which we call Jupiter. As for the land and the sea, they are under the rule of Jupiter Plutonium that nourishes mortal living beings and those that produce fruit. It is the energies of all of them that give subsistence to the earth, the fruits and the trees.»* (Treatise from Hermes to Asclepius. Verse 27) Would these bodies be just being formed and their energy already influencing our world?

וַיֹּאמֶר אֱלֹהִים תַּדְשֵׁא הָאָרֶץ דֶּשֶׁא עֵשֶׂב מַזְרִיעַ זֶרַע עֵץ פְּרִי עֹשֶׂה
פְּרִי לְמִינוֹ אֲשֶׁר זַרְעוֹ־בוֹ עַל־הָאָרֶץ וַיְהִי־כֵן׃

Genesis 1:11 adds the account of the production of trees, seeds and fruits as something that must occur in a sense of "greenness": "make the Earth green." We can read: «*ve.iamer Elohim tadeshe ha.aretz deshe esheb mazria zera etz pri oseh pri lemino asher zaró-bo al-ha.aretz ve-iehi-ken*», where there are several words that are translated differently, although they are the same or hardly varies its root. For example, it is stated that 'Tadeshe' should be made - from the form 'Deshe' (to produce, swarm) - from 'Esheb' (grass, grass, vegetation). As we can see, the cognate of 'Deshe' and 'Esheb' is similar, and something similar happens with the form 'Mazria' (with seed, seed) and 'Zera' (seed, offspring), which accompany these words in this phrase. He affirms that these seeds of "fruit trees" that in turn also produce products that also have seeds, and that all of these are of different species, depending on the type of seed, and that all this occurs on Earth.

וַתּוֹצֵא הָאָרֶץ דֶּשֶׁא עֵשֶׂב מַזְרִיעַ זֶרַע לְמִינֵהוּ וְעֵץ עֹשֶׂה־פְּרִי אֲשֶׁר
זַרְעוֹ־בוֹ לְמִינֵהוּ וַיַּרְא אֱלֹהִים כִּי־טוֹב׃

Verse 12 of Genesis 1 tells us: "*ve.totze ha.aretz deshe esheb mazrea zera laminehu ve.etz oséh-pri asher zaró-bo leminahu ve.irá Elohim ki-tob*", and we can see that first an order is given , then the order is executed, which suggests a determination that comes from "above" and that is subsequently carried out. That is, it is not something that produces the "mind of God", since, playing to this idea, why would god have to say anything, or give an order to the Earth as if the Earth could create something by itself? This means that the order was deliberately given so that in our world there would be fruit trees with their genetic information that would produce fruits that, according to their own typology, would produce more fruit trees of their corresponding species. Also, if it is something that comes from God, why does he have to "see" that it turns out to be "good" in the end? And is it that perhaps something bad was going to come from God or go wrong? What he implies is that the project that he orders to be unleashed goes well, turns out to be a success.

וַֽיְהִי־עֶ֥רֶב וַֽיְהִי־בֹ֖קֶר י֥וֹם שְׁלִישִֽׁי

Thus, with these words, Moses ends the account of the things that happened in that Third Aeon, which for us could indicate all the periods of the Paleozoic and Mesozoic era. This is not only assumed by the chronologies of the origin of Pangea and Panthalassa, but by the very fact that the most remote tree fossils have been found from periods ranging from 140 to 300 million years. Precisely the end of the Paleozoic was the Permian, about 286 million years ago, and the Mesozoic eras (Cretaceous, Jurassic and Triassic) ranged from 144 to 248 million years. This does not mean that the plants appeared immediately when the earth emerged, so however long it took for "life" to be planted here and these "experiments" were carried out successfully, it could have been throughout all of the Mesozoic. Thus life would have evolved, not alone, but through improvements made by those who responded to the guidelines of "elohim".

Day Four

Once these three eras have elapsed - since the origin of the universe of which these stories tell us - the moment seems to have arrived in which the great luminous spheres and planets began to fix themselves in specific orbits. This is how Enoch describes this development: "*And on the fourth yom I ordered that there should be the great luminaries in the circles of the Heavens*." (2nd Enoch 30:3) He speaks to us in Hebrew of 'Jugei', of the form 'Jug' (circle), but what are the "circles" of the heavens? Could it refer to the "circular" things in the heavens? If so, it could be talking about planets and stars. For contemporary astronomy, it could be considered that the "circles of the sky" could be stellar mechanisms, such as orbital movements, local voyages of stellar systems, galactic voyages of local systems, the galactic rotation itself or the movement of local galaxies. In the case of Enoch, it seems to describe the same thing as Moses, speaking of the "meorot ha.gdolim", or "great lights". However, why does it say "great lights" and not "stars" (kokabim)? The kojabim, or Kokabim, can be understood as both planets and stars, but the Meorot identify something beyond this, starting with the fact that various puns are implied here.

The word 'Meorot' is a plural of 'Meoré', which in turn derives from the form 'Aor' or 'Or' (light), which means that they are "light carriers" or "light givers". It would be more accurate to translate it as "those that provide light" or "those that are luminous". Within the phonetic concepts themselves, Meoré resembles the form 'Moréh' (teacher, professor), indicating a similarity with the idea of "guide". 2nd Ezra 6:45-46 tells us that "on the fourth day you ordained that the sun should shine and the moon should give its shine, the stars should be in order. And you gave them a burden for the service of man, which was to be done." Although it had already been made clear with the fruits that the vegetation was going to produce, now the stars complemented this work. But if the analogy between stars and angels was already understood to revolve around the Yom One and Two, would there also be a relationship here? We could think that the Earth – or "Earths" – was provided with rulers to watch over what had to be carried out, and this coincides with the references that we have observed in the processes that took place throughout the Third Yom. Ergo, it also seems to us to say directly that both the Sun, the Moon and the "stars" were "fixed" by a law. These two texts do not tell us that these bodies were created, but that they were commanded to do their "job": to illuminate. This suggests that these stars existed previously but had not started their work.

וַיֹּ֣אמֶר אֱלֹהִ֗ים יְהִ֤י מְאֹרֹת֙ בִּרְקִ֣יעַ הַשָּׁמַ֔יִם לְהַבְדִּ֕יל בֵּ֥ין הַיּ֖וֹם וּבֵ֣ין
הַלָּ֑יְלָה וְהָי֤וּ לְאֹתֹת֙ וּלְמ֣וֹעֲדִ֔ים וּלְיָמִ֖ים וְשָׁנִֽים׃

Moses wrote in Genesis 1:14, "*ve.iamer Elohim iehi meorot ba.rakia ha.shamaim lehabdil bein ha.laila ve.haiu leotot ve.lemoadim ve.leiamim ve.shanim*", and if we understand that time and space revolving around the existence of masses, this passage fits perfectly, as it maintains that it was said that there must be Meorot in the firmament, or space, of "the" heavens – speaking in the plural – in order to "separate" the Laila (night) and being "signs" for the "iamim" (days, cycles, eras, eons) and "shanim" (years). The positions of these bodies would "split" the darkness and define the cycles of time. But why does it say "rakia ha.shamayim" (firmament of the heavens)? If there is an upper sky and another defined as our atmosphere, what is the rakiah of those shamaim? Definitely the space - or "empty" - that is in all the heavens, which reinforces the already accepted idea that the sky is full of stellar systems, and re-emphasizes that the heavens are places that encompass infinity.

וְהָי֤וּ לִמְאוֹרֹת֙ בִּרְקִ֣יעַ הַשָּׁמַ֔יִם לְהָאִ֖יר עַל־הָאָ֑רֶץ וַֽיְהִי־כֵֽן׃

In the story of Moses, all the celestial bodies seem to be included as 'Meorot', so the 'Shemesh' (sun), the 'Iareja' (Moon) and the 'Kojabim' (stars, planets) enter the description. Now, is this pattern repeated throughout the universe or only in our solar system? Being methodical in the analysis, we must consider that the name 'Sun' is Nordic, while 'Shemesh' derives from the Akkadian 'Shamash', and this from the Sumerian 'Utu' (the Greeks called it 'Helios', and they were not very wrong, since the sun supposedly stays incandescent due to the fusion of helium). The word 'Shamash' or 'Shemesh' means "identifies fire" or "who has the name of fire", even Enoch says that it also receives the name of 'Ur-Jamah' ('burning light' or 'hot light'). For its part, the Moon is called 'Iareja' or 'Iareaj' because of the definition of the "month", while according to Enoch it also has the name of "Lebaná" (white), and it is called Iareaj because it defines the cycles of 30 days that separate the year into 12 parts. So the Sun sets the 12 hours of the day and the time of a total day (24 hours), the Moon with its phases the stages of the month, while the other stars define the other cycles: the planets of our system mark the eras and stages. more detailed ones of longer and encrypted cycles, while the stellar groups that we call 'constellations' define the others, and in addition, the even longer eras and the total journey of the sun, or the complete solar cycle.

However, the stars define the time, but they also illuminate the Earth. 2 Enoch 30:4 says: «*In the circle, the first, the highest I placed the star Shabetai, in the second above I placed Noga, in the third Maedim, Shemesh fourth, Tzedek fifth, Kojab sixth, seventh Iareach; and in the stars, the small ones, the splendid ones, those are the aerial ones, the last-final ones.*» These are the Hebrew names for the planets in our solar system up to 1780, when Uranus was discovered, and then Neptune in 1846. Shabetai is Saturn; Noga is Venus; Maedim is Mars; Shemesh is Sun; Tzedek is Jupiter; Kojab is a star and, likewise, the name given to Mercury; Iareaj is Moon. This order that Henoc gives is strange, since it does not correspond to the sequence of sizes of these bodies or their location in the solar system, so it could refer to their initial state.

What is evident is that if we previously asked ourselves about those "circles" of the heavens, these 7 space bodies were distributed from largest to smallest in that order. In addition to the fact that nothing is said about Uranus and Neptune - at least speaking of that era of the creation of the solar system - mention is made of other "stars" or "planets" that are "small" - logically in relation to the previous ones -, described as "splendid" who are "at the end", or who are the "last".

It is not clear if he is talking about exo-planets, dwarf planets of the solar system and/or natural satellites, but he says that they are from 'ha.Avir' (from the air). It would not make sense for him to be insinuating that there are planets or small stars in our atmosphere, but if you remember the theses in 'The Rebellion of Sakla III' and 'Transcendence', you will be able to collect the data on the "regions of the air", which is the way to refer to the outer space closest to our orb, that is, what to a certain extent we could define as the first heavens of the Earth. The ancient peoples referred to the space of the stellar Sphere as the regions of the air, and that we could say that it is everything that, including the Earth, encompasses the space that reaches beyond the 12 constellations of the zodiac and the 36 decans.

Whether it refers to bodies in this system, or to the constellations (which I would define as "small" because of the distance - being an honestly ridiculous conclusion -), the fact is that it encompasses all these stars. The mention of Moses in Jubilees 2:8-11 does not give too much detail either: "*And on the fourth day he created the sun and the moon and the stars, and set them in the expanse of the heavens, to give light to all the earth, and in the day and the night, and divide the light from the darkness. And God appointed the sun to be a good sign on earth for days and for Sabbaths and for months and for celebrations and for years and years of Sabbaths and for anniversaries and for all seasons of the year. And that separates light from darkness [and] from prosperity, that all things may prosper and grow soaring upon the earth. These three guys made the fourth day.*"

וַיַּעַשׂ אֱלֹהִים אֶת־שְׁנֵי הַמְּאֹרֹת הַגְּדֹלִים אֶת־הַ מָּאוֹר הַגָּדֹל לְמֶמְשֶׁלֶת
הַיּוֹם וְאֶת־הַמָּאוֹר הַקָּ טֹן לְמֶמְשֶׁלֶת הַלַּיְלָה וְאֵת הַכּוֹכָבִים:

Verse 16 of Genesis 1 says, «*ve.iás Elohim et-shnei ha.morot ha.gdolim et ha.maor ha.gadol lememshelet ha-iom ve.et-ha.meor ha.katan lememshelet ha.laila ve.et ha.kokabim*», quote that says that the greater Meorá has the objective of dominating or ruling over the Yom (aeon, day), while the smaller one has it over the Lalila (night). These two are the key Meorot, and then there are the kokabim, which, as we have seen, define the rest of the meorot, be they planets or stars. As can be seen in the following quote that says «*ve.iten otem Elohim ba.rakia ha.shamaim lehair al-ha.aretz*», we observe that they were designated to "illuminate" on the Earth, and just as in the text of Enoch, we find the word 'Air', but interpreted as the conjugation of the form 'Aor' (light), in the sense of illuminating, so it would have a double meaning: region of the air and illuminating (let us bear in mind that the stars and the Moon they illuminate the night sky - especially Selene due to its proximity and consequently its volume -, although not the Earth, but the Sun does, since given its proximity the photons hit the entire atmosphere illuminating it, in such a way that they produce the effect of "light of the day"). Moses says, then, in Hebrew characters, this one in verse 17:

וַיִּתֵּן אֹתָם אֱלֹהִים בִּרְקִיעַ הַשָּׁמָיִם לְהָאִיר עַל־הָאָרֶץ

2 Enoch 30, for its part, adds: «*And guard over the Heavens to illuminate the Day and over the Moon and over the stars to illuminate the Night. And the Sun must go around the whole circle [of the constellations] of the zodiac; and 12 circle the zodiac revolve [around] the Moon; and they act according to the meaning of their names and thunder according to the circle of the zodiac that is in front of them, and [the] law of their hours according to their rotation.*" (Vers. 5-6) It is fascinating to see that it was already known at that time that the sun makes a trip around the constellations, since it is a concept that would be difficult to know without a computerized system, since it consists of a cycle of 2,150 years for each house, being a

total of about 25,800 years (ironically this is the total time with respect to the precession of the equinoxes, and the cycle that many consider to correspond to the journey of the Sun around Alcyone, which would be the star center of the Pleiades, being the sun the 8th of said group). As in the previous case, Moses reiterates in chapter 1:18 that the role of these star kings is to dominate in the Yom and in the Laila, and to separate between Light and Darkness, seeing it as something positive:

וְלִמְשֹׁל֙ בַּיּ֣וֹם וּבַלַּ֔יְלָה וּֽלֲהַבְדִּ֔יל בֵּ֥ין הָא֖וֹר וּבֵ֣ין הַחֹ֑שֶׁךְ וַיַּ֥רְא אֱלֹהִ֖ים כִּי־טֽוֹב

Enoch's descriptions of the 'Meorot' are clearly an astronomy treatise thousands of years old: "*Observe all the things that happen in Heaven, how the luminaries of Heaven do not change their path in the positions of their lights and how they are all born and put (install), ordered each one according to its season and do not disobey its order.*» (1st Enoch 2:1) The positions of its lights? According to what this legendary prophet tells us, the Meorot are established in specific places, not in a random order, and their light intensity (magnitude) is deliberately set according to a pattern or prerequisite. In addition to this, Enoc speaks of the Meorot as if they were "consciousnesses", not mere spheres of light (assuming that we are also talking about 'spheres' here), and that they definitely participate in the reality of existences and even in determinations. universals and the right of free will: «Rauel» is «*one [of] the malachim (messengers), sacred, [who] takes revenge on the world [of] the luminaries*».

If the Meorot had no conscience or conscious actions, how could they do something that would lead the archangel Rauel to "take revenge" on them, or retaliate? It is clear that the close relationship between the idea of Meorot (luminary, light) and Kokabim (star) with that of Malajim (angel) is so close that it even seems to indicate that these stars are a high manifestation of consciousness and/or a consciousness in around which conscious beings are connected and linked. For those who doubt this relationship, and even the fact that the "heavens" are indeed portions of the universe in relation to planes and dimensions that

are intertwined with the physical, psychic and ethereal, Henoc has some very constructive words: «*The angel Michael He took me by the right hand, lifted me up and led me into all the mysteries (secrets) and revealed to me [...]*
...the secrets of the limits of Heaven and all the reservoirs of the stars, of the luminaries, whereby they are born in the presence of of the saints He transferred my spirit into the Heaven of Heavens and I saw that there was a crystal building and between those crystals, tongues of living fire.» (1st Enoch 71:3-5) Not only does he make a difference between Kokabim and Meorot again, but he says that the angels are in front of these "consciences" that are created, understood in the place where they are created. Where are stars and planets created? Unless they are produced in another universe and spit out to this one, it seems obvious that there is an interaction of worlds, and this takes place in "the heavens" of some perceptible room. Another even more efficient explanation is that, as the Oahspe explains, there are stars that have double composition and have been produced from the ethereal worlds to illuminate and awaken consciousness in the physical worlds. Oahspe would also clarify that having 3 spheres of reality (corporeal, atmospheric or psychic, and ethereal), this type of space bodies would develop between these levels (being the nebulae – the cradle of formation of these stars – of an atmospheric, or psychic nature).
In the 'Book of the Luminaries of Heaven', Enoch describes each sequence, cycle and period, both of the Sun and the Moon, and the establishment they have and the angels that direct them to fulfill their objective: «*The Book of Movement of the Celestial Luminaries, the relations between them, according to their class, their dominion and their station, each one according to its name and the place of its departure and according to its months, which Uriel, the holy malaj who was with me and who he is your guide, he showed me and revealed all his laws exactly as they are and how they are observed all the years of the world, until eternity, until the completion of the new creation that will last until eternity. This is the*

first law of the luminaries, the luminary of the sun, which has its birth at the eastern gates of Heaven and its setting at the western gates of Heaven. I saw 6 gates where the sun rises and 6 gates where the sun sets, and the moon rises and sets through those gates, as well as the leaders of the stars and those who guide them." (1st Enoch 72:1-3)

Then he adds about this "Greater Light": «*First there appeared the great luminary whose name is Shemesh (the sun) and whose circumference is like the circumference of Heaven and is totally filled with a fire that illuminates and burns. The Ruach carries the chariot in which he ascends and the sun sets and returns through the north to return to the east and is driven to enter that gate and shine on the face of Heaven.*" If the Greater Light has a "circumference" that is equivalent to the "circumference of heaven", this coincides with Zephaniah's and Baruch's revelation about the journey of these knights to the heavens, since they define such heavens as "spaces in space", but in several dimensions, and affirm that the "distance" from the Earth to the Sun is like the size of that sky. But, what parameter would determine the magnitude or proportions of each sky?

Apparently, the planetary sphere or star itself. Oahspe calls the ethereal bodies 'Photospheres', and equally agrees that planets, stars or suns and photospheres – or vice versa – are the axis around which dimensions wrap, one on top of the other, creating what could imagined as a layer or cloak so thick that seen with spiritual eyes at a great distance, the planetary sphere itself would be almost like a minuscule dot next to its corresponding sky (or skies within its sky). The role of the Sun, like that of the Moon, would be, in effect, to counteract invisible, energetic and potential forces, preventing the event and darkness, with all their shadowy powers, from making balance and order disappear.

For this reason we see descriptions as apparently incoherent as saying that the sun watches over the day and the night, since he is the one who determines this. But if we understand that Day and Night are concepts, it makes total sense that the Sun and the Moon have their strategic and decisive role in this balance: «*This is the law of the path of the sun and its return, according to which it returns and is born 60 times, so the great light that is called the sun, forever and ever. The one who rises is the great luminary, named after its own appearance, as the Lord has commanded. Just as it is born, it hides, without decreasing or resting, but traveling day and night; and its light shines 7 times more than that of the moon, although when observing them both they have the same magnitude.»* (1st Enoch 72:35-37)

Finally, and in order not to go on much longer, I will comment that the Moon is a reference "satellite" with respect to the constellations of the zodiac and the Earth, an electromagnetic influence for the energy fields of the Earth, and all this, in relation to the control of the powers and forces that govern chaos (since the magnetic, photonic and gravitational force - being related - are blocks for "doors" or portals that open and close in their steps, convergences, alignments, shadows (eclipses) and movement, like a great gear clock, to allow the transit of one or other powers of the invisible planes of reality that maintain the balance of the cosmos): *«After this law, I saw another law, which deals with the small luminary , whose name is moon. His circumference is like the circumference of Heaven and the chariot in which he rides and the light is given to him in moderation; and each month its birth and setting are modified; its days are like the days of the sun and when its light is full, it is the seventh part of the light of the sun.»* (Chap. 73:1-3)

If the Moon also has a circumference similar to that of the sky, it means that it belongs to the same sky that is being talked about or, more probably, it has a field of extension that in proportion follows the same sequence of the spiritual worlds that are on the higher planes of the "physical" spheres. An example of all this can be taken from the Pistis-Sofia treatise, where Valentino allows us to know the words of the Lord Jesus about the celestial battles: *«And their angels, and their aeons, and their archangels, and their archons, and their gods, and their lords, and their forces, and their luminaries, and their ancestors, and their triple powers, saw that I was infinite light, to which no kind of light is alien.»* (Chap. 3:20)

In this manuscript from the Nag Hammadi Library it is read that Jesus describes all the spaces of consciousness and the envelope of spatial bodies as 'spheres', and also speaks of the "firmament" as a spherical space that is limited at the height of what could be interpreted as the limit of this solar system; then, another sphere would be the one that encompasses the previous one, being something like a space where other stellar systems reside, and on this, another sphere, which would be the circle of the zodiac constellations. In addition to speaking of Mars, Mercury, Venus, Jupiter and Saturn as consciousnesses that are interrelated with the 12 constellations and the 36 deans, Jesus in this text is very emphatic in stating that all these celestial bodies are beacons or starting points of the spheres of existence and destiny, everything being operated from different dimensions: «*And it would have been a long time before the archons of the aeons, and the archons of Fate, and of the sphere, and all its regions, and its heavens, and their aeons, have been destroyed.*" (Ev. Valentine 5:7)

Day Five

If in itself the appearance of plant life is fascinating, how much more is animal life? Accepting that plant eukaryotic cells could not appear alone, it is conceivable that the same thing also occurred with animals. The prophet Enoch speaks to us about this in just one verse, saying: «*And on the Fifth Day I drew out of the sea and there came out fish and birds of various kinds and every reptile that crawls on the Earth and what goes on four on the Earth and flies in spirit male and female in the midst of them, and every soul that breathes for all life.*" (2nd Enoch 30:8)
According to ancient texts – and modern science seems to agree with them – the first forms of life in our world were androgynous creatures, that is, hermaphrodites, each possessing both sexes: male and female, at once. This 5th era, which seems to be the general period of the Mesozoic, saw aquatic, aerial and "reptilian" life.

It would seem that this supports the hypothesis of evolution, but what this seems to adduce is that birds appeared before reptiles, or that they would have arisen "out of the water", which does not have to be correct. Enoch wrote that he was told by the Creator that life was taken from the sea in this period, and according to what was carried out, it was successfully achieved that such life forms were produced in large numbers in the ocean - and /or in fresh water -. What also seems to imply is that the same project managed to create "birds". The term 'Aof', although it is translated as 'bird', gives rise to the word 'Meofef' (to fly), in the sense of the action or exercise carried out by birds: flight. Genesis tells us that the reptiles appeared on the Sixth Yom, so the birds could not have come from the reptiles. Likewise, the texts maintain that the first forms of life came out of the "sea", so it is to be assumed that what they intend to imply is that the life "experiments" were carried out offshore, and resulted in the forms of life marine and, later, those that would cross the skies (first androgynous and then already separated by complementary sexes one from the other for the reproduction and maintenance of life).

In the case of Moses, both in Genesis and in Jubilees, this account is more explicit: «*And on the fifth day he created great monsters of the sea in the depths of the waters, for these were the first things of the flesh that were they have created by their hands, the fish and everything that moves in the waters, and everything that flies, birds and all their kinds. And the sun rose above them to prosper (them) and, over all that was on the earth, all that sprouted from the earth, and all fruit trees, and all flesh. These three types He created the fifth day.*" (Jubilees 2:11-13) Here he tells us about something that Enoch deals with in another section, and it is about those "sea monsters" that the Septuagint translates from Genesis 1:21 as 'Kiti'. From the Greek word 'Kiti' comes 'Kitei', and from there 'Kiteos', which derived from the Latin 'Cete' and 'Cetus', and into Spanish as 'Cetacean'. It is common to hear in the stories of ufology that beings from other worlds brought life to Earth, being the first creatures with consciousness - which according to the defenders of these theses they would have brought from Sirius - the cetaceans: dolphins and whales.

Another aspect is the one that encompasses in these creatures the character of giant reptiles. The word that Moses uses in Genesis is 'Taninim', which is usually understood as 'dragons'. What is clear is that these affirmations coincide in saying that these forms of life were the first of "meat" that came to be produced; It also includes the marine biodiversity of so many fish that exist, and that were also created, as well as the rest of the things that live in the sea. This passage from Jubilees is also important, because it makes it clear that those "flyers" were not simply birds, but "zbubim" (flies, flying insects). We had also said that plant life was produced on the Third Yom, but that the sun did not enter the scene until the following cycle (the Fourth), and this is where it says that then! The sun crossed the sky and made everything "prosper".

So that the plants and trees, despite existing in the past, only really began to bear fruit from this moment, so that before they must have been in some kind of "vegetative" or torpor state, without yet producing fruit. Consequently, it is evident that it was not seeds that were disseminated on the ground, but vegetation already prepared -previously arranged- for when the passage of the sun arrived. This suggests whether the plant life of that time was far from resembling the current one and having a more long-lived and quasi-immortal appearance.

וַיֹּאמֶר אֱלֹהִים יִשְׁרְצוּ הַמַּיִם שֶׁרֶץ נֶפֶשׁ חַיָּה וְעוֹף יְעוֹפֵף עַל־הָאָרֶץ
עַל־פְּנֵי רְקִיעַ הַשָּׁמָיִם׃

In Genesis 1:20 we read: «*ve.iamer Elohim ishratzu ha.maim sheretz nefesh jaiah ve.of iofef al-ha.aretz al-pnei rakia ha.shamaim*», which translated means that the sea was said to “run” a swarm of "living souls" and flying birds on the earth, in front of the celestial firmament. It is important to understand that all these words and phrases have a double representative component, since in addition to its direct meaning it has a symbolic nuance, wanting to show an analogy with the work of angels. Thus, all things in the world of phenomena are images of heavenly things. However, these "living souls" had not yet been described, and they come to be what in the Hebrew language is 'Nefesh Jaiah', that is, a form of life that depends on breathing (air), which has a soul, the privilege of existence and movement in a sphere of reality (in his case, the physical world).

Like Enoch, Ezra speaks of these "monsters" and the forms of life that were created in that era: «*On the fifth day you said that in the seventh part, where the waters were gathered, that there should [be] living beings, birds and fish, and so it happened. For the dumb and lifeless water gave birth to living beings at the commandment of God, [so] that all people could praise your wonders. So, you ordered two living beings, one you called [Be]he[m]o[t], and the other Leviathan; And you have separated one from the other: from the seventh part, that is, where the water came together, it could not support both. And to [Be]he[m]o[t] you gave a part, which dried up on the third day, that they should live together in the same part, in which are the thousand hills: However, to Leviathan you gave the seventh part, that is, humidity, and you have kept it to be a devourer of whomever you want, and when.*" (2nd Ezra 6:47-52)

Unless Behemoth and Leviathan were reptilian mega-whales, these creatures don't seem like the norm. Making an in-depth study on this matter, one could conjecture that these two "beasts" constitute a set of "creatures" under one spirit. For this reason, speaking of the end times, Henoc was told that "*on that day two monsters will be made to come out separately, one female and one male. The female monster is called Leviathan and lives at the bottom of the [great] sea above the source[s] of the waters. The male monster is called Behemoth, it rests on her chest in the immense desert called Dondain, to the east of the garden inhabited by the elect and the just, where my old father was taken, the seventh since Adam, the first man created by the Lord of the spirits.*" (1st Enoch 60:7-8)

If these two monsters are reserved for the end of time, why does the Apocalypse of John not speak of them? According to Enoch, these beasts will devour meat, the writer implying that this "meat" is human. Apocalypse the only thing that speaks in analogy with something like this is regarding two powers that will subdue the world for a few years. If we look at the similarities - as can be seen in the 'Sakla Rebellion' books – Behemoth (which means "the beast") is a male creature that, according to Jewish tradition, rules over all land animals; For its part, Leviathan, is a female creature that, according to the same tradition, directs aquatic life forms. The most coherent correspondence is that Behemoth could be related to the Beast of Apocalypse, and Leviathan to the Dragon. The question would be, how come these creatures were created so long ago, especially if their function is for the "final" age?

The Vedic culture of India recounts in its sacred writings that at the beginning of time, when the 'Lilá' (or 'Lika') began, the Creator of the cosmos, Brahma, faced the sovereign powers of chaos, such as Vala and

Vritra, the first demon-gods, whom he defeated, but also - manifested through Vishnu - fought against the titanic Rajú (the 'Attacker'), and divided this creature into two: Rajú and Ketú (the one who was left "disconnected" and wandering through space). The attacker is called 'the Proud One' in ancient Hebrew history, and it is told of him in the psalms: «*You broke Rahab as one mortally wounded; with your mighty arm you scattered your enemies.*" (Ps. 89:10, R95) This victory is also mentioned in the books of the prophets: «*Wake up, wake up, clothe yourself with power, arm of Yaheveh! Wake up like in the ancient time, in the past centuries! Aren't you the one who tore Rahab to pieces, the one who struck the dragon?*" (Isa. 51:9)

If we speak here of celestial battles, we have to add to our analysis the fact that the Enuma-Elish relates that the "evil spirits" that were on "the Earth" - or solar system - in the primeval era, were forced to to split, becoming one of two parts that survived an astronomical impact (reminiscent of a war between supernatural beings). Those that were thrown into the world and those that were thrown into space came to be considered the two spirits of evil that had to be kept apart, since reunited again, they would procreate chaos. So, following the narrative, we find that these creatures were created on the Fifth Yom, and since it does not tell us when they were separated, we have to assume that it was in that same cycle. Therefore, it could be conjectured that at that time some kind of war would have occurred already in physical-psychic canons, not between humans (who did not yet exist), but between "intelligent" beings.

Several of the Nag Hammadi manuscripts speak of the celestial wars and their seasons, saying, for example, that time after chaos and its powers had appeared, there were various battles in these heavens for their sovereignty, and the evil powers were divided and subdued. The great similarity between these accounts makes one imagine a somewhat fantastic, but ordered story of how various conflicts were taking place in the heavens, and also the Earth was experiencing them, even at the time when organic life and oviparous creatures were created on Earth. our world (and possibly the war issue was related to the plans for the creation of life and conditioning of this planet).

וַיִּבְרָא אֱלֹהִים אֶת־הַתַּנִּינִם הַגְּדֹלִים וְאֵת כָּ ל־נֶפֶשׁ הַחַיָּה ׀ הָרֹמֶשֶׂת
אֲשֶׁר שָׁרְצוּ הַמַּיִם לְמִינֵהֶם וְאֵת כָּל־עוֹף כָּנָף לְמִינֵהוּ וַיַּרְא אֱלֹהִים
כִּי־טוֹב׃

Moses recounts in verse 21 of Genesis 1: «*ve.ibrá Elohim et-ha.taninim ha.gdolim ve.et kal-nefesh ha.jaiah; ha remeshet asher shertzó ha.maim laminehem ve.et kal-aof kanaf laminehu ve.ira Elohim ki-tob*» , speaking of "creating" great dragons and every living soul. Sometimes he says to "create", others that he commands to "produce" and others that a certain thing "comes out", each case being different from one another, since on this occasion it is clear that the gods (elohim) create, while in the other situations they "command" - or who is over them - that "something" or "someone" carry out functions of "producing" based on the elements already created, or simply that the elements themselves, or existing mass, it is transmuted to establish new scenarios. Now, why would he create "big" dragons? It is notorious that he refers to the fact that he created two sides, to lessen the power of these forces, but he also speaks on the earthly side of the creation of great dragons.

Celtic records tell us, saying that the "Earth was clothed in the mantle of the lady of green, grass covered the face of the earth. The waters produced fish and the creatures that move about and writhe and coil in the waters, the serpents and terrible-looking beasts that were of old, and the creeping and crawling reptiles. There were things stomping around, and dragons in hideous form cloaked in terror, whose great bones can still be seen. Then came out of the belly of the earth all the beasts of the field and the forest." (The Book of Creation. The Kolbrin). Unequivocally these are direct and indirect allusions to dinosaurs. In Genesis a difference is made – or so it seems – between two groups: taninim ha.gdolim (great dragons) and nefesh ha.jaiah (living creatures with souls). Moreover,

these living souls are defined as "all", that is, on the one hand there were these mega reptiles, and on the other, "all" living souls.
Does it mean that the other big monsters weren't creatures with souls? In fact, these living souls are categorized as those first oviparous, namely, forms that crawl on the subsoil or sea bed, and the "winged" birds. But are there birds that are not winged? If you misunderstand that a characteristic of birds is their wings, why do you highlight it, and why didn't you do it before? It is probably because it speaks of an order of creation, where fish and INSECTS (the simplest complex life forms) were produced first, and just after "laboratory" work developed dinosaurs, molluscs and crustaceans, and the birds. It might be thought that the vast jungle expanse of the mainland was so depopulated by animals that testing the design of the first reptiles there would be ideal, although due to its size it got out of hand (the other forms of life that would accompany them - birds – they would be flying, out of reach).

וַיְבָ֧רֶךְ אֹתָ֛ם אֱלֹהִ֖ים לֵאמֹ֑ר פְּר֣וּ וּרְב֗וּ וּמִלְא֤וּ אֶת־הַמַּ֙יִם֙ בַּיַּמִּ֔ים וְהָע֖וֹף
יִ֥רֶב בָּאָֽרֶץ׃

Genesis 1:22 says: "*ve.ibraj otam elohim leemor pru ve.rabú ve.malú et-ha.maim ba.iamim ve.ha.aof ireb ba.aretz*", referring to the fact that these forms of life had to fructify and multiply in the sea, while the birds did it on Earth. But why only birds? Aren't dinosaurs supposed to be multiplying on the dry surface, too? This makes one wonder if the great dragons to which he initially refers were indeed just the oceanic monsters, such as megalodons, plesiosaurs and other species that were the giant companions of whales and other life forms that are still studied in cryptozoology, such as giant squids and others spoken of in Hindu and Japanese myths. The stories and engravings on these specimens are seen in many places, and it is known that their size was colossal, mixing with the stories about the fights between the gods, such as when Horus fought with Set (in Egyptian stories), which is the myth Greek of the confrontation between Zeus and Typhon.

In his book 'Flying Serpents & Dragons' (1990), RA Boulay theorizes on the direct relationship between the appearance of dinosaurs and the wars between reptilian-like demonic beings, using all kinds of sources, including the Bible and the Hagadah itself. (a text from Jewish tradition). The use of analogies is key to understanding both levels of information received, as I have already said, since the messages are given with similarities between the celestial and the earthly, showing the etheric and the phenomenal as one in the image of the other. Other characters in Hebrew history say that these factions represented the thousand-year-old dragon, some as 'Fugitive Serpent Leviathan' and others as 'Torturous Serpent Leviathan' (Isa. 27).

Yeshua (Jesus) tells about the celestial combats precisely describing them as battles against dragons, which is a classic that appears in practically all prehistoric tales or fables: «*And when the power of the triple power had descended into chaos, he found Pistis Sofia. And the force with the face of a lion, and the force with the face of a serpent, and the force with the face of a basilisk, and the force with the face of a dragon, and all the forces of the triple power surrounded Pistis Sophia, wanting to snatch from her a second time her powers. forces. And when they tormented and afflicted her, she turned again to the light.*" (Ev. Valentino 20:10-12)

The account of the Enuma-Elish (Sumerian epic of creation) makes us know all this in many details, but clearly through figurative descriptions: «*They themselves [were] gathered together and at the same time advancing Tiamat side; They were furious; They played mischief without rest Day and Night. They prepared for battle, fuming and fury; They joined forces and made war, Umm-HUBUR [Tiamat] Maker of all, Made into invincible weapons of addition; she engendered serpent-monsters, sharp-toothed, and merciless [...] With poison instead of blood, she filled their bodies with ferocious snake-monsters, dressed in terror. With the splendor that dressed them as a gala, he made them of high stature...*»

The tablet goes on to say: "*Whoever contemplated [at] them, terror seized him, their bodies reared up and no one could resist their attack. He established snakes and dragons, and the monster Lahamu, and hurricanes, and mad dogs and scorpion-men, and strong storms, and fish-men and rams. They carried cruel weapons, fearing no fight; His commands were powerful, he couldn't resist them; before this appearance, of great stature, he made eleven [kinds of] monsters. Among the gods that were her children, insofar as she had given her support, she exalted [a] Kingu; in their midst he raised to power. To march before the forces, to lead by host, to give the battle signal, to advance to the aggression, to lead the battle, to control the fight, To him he entrusted; in expensive dresses that she made him sit on..."*

This Sumerian description encompasses all of the above, speaking of what appear to be demonic creatures of different species, all of them participating in a war in the solar system, since Lahmu, Kingu, and other deities described throughout the conflict, are the names received by the planets and moons of our solar system in Mesopotamia.

Days Six

The last aeon of creation is the 6th, the stage in which terrestrial reptiles and mammals appear, and also the human being, the last to manifest and for whom it seems that everything was arranged and conditioned. About this period Ezra wrote: «*On the sixth day you gave a commandment to the earth, that it give beasts, cattle and reptiles, before and after these, also Adam (man), whom you made lord of all your creatures. From it all come, and also the people you have chosen. I have spoken all this before you, O Lord, because you made the world out of love for us.*" (2nd Ezra 6:53-55) In accordance with these words, the order was given to create mammals of two genera and reptiles - previously, and deliberately, the arrival of man -, and concluding with another reaffirmation that everything was established for the human to be lord over everything and dispose of everything.

Jubilees contains: «*And on the sixth day He created all the animals of the earth, and all the cattle, and everything that moves on the earth. And after all this he created man, a man and a woman he created them, and gave him dominion over everything that is on the earth, and in the seas, and over everything that flies, and more beasts and more cattle, and over everything that moves on the earth, and over all the earth, and more than all this he has given dominion. And these four guys that he created on the sixth day. And there were twenty-two guys altogether. He and all his work [was] finished on the sixth day; what is in the heavens and on the earth, and in the sea and in the abysses, and in the light and in the darkness, and, in everything.*" (Ch. 2:14-16)

Here he speaks to us of the 'Chaiat' of the Earth, of 'Behemat' and of the rest of what "moves", possibly meaning "what crawls" (in Hebrew 'ha.Remeshet'). The word 'Jaiat' (living) is the conjugation of 'Jaiah' (alive), whose root is 'Jiah' (life), which is why it is understood as a living "being". Behemat is a synonym, more explicit according to the animal or animalistic character, but which is used to define beasts and, to a greater extent, for cattle (hence, some interpret it as a domestic or pack animal). It is very rare that he always makes a distinction between 'Jaiat' or 'Behemat' to include terrestrial animals (with the exception of reptiles, and, by extension, it is understood that amphibians as well).

This curiosity suggests that 'Jaiah' would be the equivalent of the Latin 'animam' (animal, animated being), while 'Behemah' would be the equivalent of the Latin 'bestia' (wild animal, irrational creature), which includes wild animals that they have become domesticated. For our society, a generality as simple as saying "animals" would have been understood more than enough, so why emphasize this as if they are two different sets? Possibly because it is characteristic of all ancestral culture to know how to differentiate between coexistence animals and dangerous animals, as well as carnivorous creatures and herbivorous (or omnivorous) creatures, without counting the euphemisms to refer to intelligent beings, due to their attitude or virtues/defects.

וַיֹּאמֶר אֱלֹהִים תּוֹצֵא הָאָרֶץ נֶפֶשׁ חַיָּה לְמִינָהּ בְּהֵמָה וָרֶמֶשׂ
וְחַיְתוֹ־אֶרֶץ לְמִינָהּ וַיְהִי־כֵן

Moses says in Genesis 1:24, "*ve.iamer elohim totze ha.aretz nefesh jaiah laminah behemah ve.remesh ve.chaito-aretz laminoh ve.iehi-ken*" as in the other passages, he tells us that they occur in the Earth 'nefesh jaiah', which he defines as 'behemah' according to all its species, 'remesh', of all its species, and 'chaito-aretz', according to its species. Some have conjectured that this reference makes a distinction between the levels of consciousness and development of these creatures, and/or greater differences, such as the very existence of other "beings" of an intelligent nature, along with reptiles and beasts. Seen analogically with the spiritual, it could be assumed that, as well as animal forms, there were 3 types of beings with higher consciousness than animals, whether in this dimension or in others in our sphere.

This conjecture is reinforced by the study of ancient manuscripts that speak of "mysterious" beings that coexist with wild forms of life, but live - or lived - in areas far from the population centers of civilization. In his case, NH II, 5 (Treatise on the Origin of the World, from Nag Hammadi), says that «*all plants germinate on earth according to their kind, with the seed of authorities and their angels. Then, from the waters, the authorities created the animals, if any, and reptiles and birds according to their kind, with seeds from the authorities and their angels.*" (Vers. 21-28) The reference to “seed” encompasses both vegetable grain and “seed” or “semen” in an animal reproductive biology sense, which may suggest that these authorities and their angels used their own genetics. for the elaboration of life on this planet.

וַיַּעַשׂ אֱלֹהִים אֶת־חַיַּת הָאָרֶץ לְמִינָהּ וְאֶת־הַבְּהֵמָה לְמִינָהּ וְאֵת
כָּל־רֶמֶשׂ הָאֲדָמָה לְמִינֵהוּ וַיַּרְא אֱלֹהִים כִּי־טוֹב׃

Genesis 1:25 tells us, "*ve.iás elohim et-chaiát ha.áretz lamináh ve.et-behemáh laminah ve.et kal.remésh ha.adamáh laminéhu ve-irá elohim ki-tob*," highlighting that the determination was made correctly and had a positive result. However, here he begins to use a word that until now he had not used: 'Adamah'. This word does not make sense to apply it if man did not yet exist, since said designation defines the name of our world because of humanity. Adamah is the feminine form of 'Adam' (man), and its root stems from the structure that formed the Coptic and Koine word 'Adamas' (diamond). This can only suggest one of two things: either there were humans on this world since then, or our orb was already "baptized" or "founded" to welcome the human race.

Strangely, Enoch is the only one of these sources that does not agree that the terrestrial animals were created or produced at the beginning of the Sixth Yom, but instead includes them all in the Fiveth Yom, possibly towards the end of this and the beginning of the Sixth. What is clear in all the narrations is that the man was the last to appear, and things basically followed the same sequence, which is what we have followed throughout these chapters; Moreover, with the appearance of man, creation was considered concluded, sealing the cycle of formation of the cosmos as a complete and perfect setting for the human experience.

Apparent gap or change in said version mean? Enoch is the one who had given more details about the creation of things, and he is much earlier than the other narrators, so what would be the reason why the deity who spoke with Enoch told him that the animals had been produced in the yom five, but man was created in the yom Six, without the animals? We can give interpretations, but the truth is that Enoch not only says this, but also adds the "process" and "patterns" that led to the appearance of the human: «*And on the Sixth Day I ordered my Wisdom to create man from 6 bases: his meat from the Adamah; their blood from "the speed [of] the malachim and the densities-clouds their tendons-ligaments; and his hair from the grass of the Adamah; his soul of my breath and my spirit.*" (2 Enoch 30:10 (Hebrew scroll 11:58)).

Could there be a relationship between the Jaiah, Remesh and Behemah with the prelude to the appearance of man in a biological sense? I'm not talking about sporadic evolution, but about deliberate genetic design: Intelligent Design. Was it the Wisdom of the deity that was behind the creation of man? As also stated in the Nag Hammadi manuscripts, there seems to be a spiritual collective representing a celestial kingdom, called 'Jokma' or 'Jajmah' (wisdom, knowledge, cunning), which directed the operations of the Holy Spirit through of a possible civilization - or a group of them - of "angels" or superhuman beings that represent God and personify the Wisdom of these stellar systems. In Amerindian cultures it is believed that there were several "eras" before the one that constituted the human - and the one prior to it - and man already existed, but had been turned into an ape (the Ehecatonatiuh era of Aztec sources).

This humanity would have been a design prior to the modern human body, and is described as "monkey" in Celtic records, who also tell -

like other peoples - that this simian human race was destroyed with the fall of an asteroid. According to ancient records, before or during the appearance of the human being, there were giants in our world, and according to Aztec records, they were in the first age, that of Nahui-Océlotl, which was preceded by the Nahui-Ehécatl era, devastated by a "hurricane" or mega disaster that supposedly turned its survivors into "monkeys". The Celtic Book of Creation, in the Kolbrin records, "*We are told how the Selok monkey-tribe, led by heavenly men, perished in flames before the Valley of Lod; Only one monkey made it to the upper heights [of the] cave. When the celestial man was reborn from the monkey in the cave of Sorrow, he could taste the fruits of the earth, and drink its waters, and feel the coolness of its winds [...] Man, created from earthly substance alone, he could not simply know the things of the Earth, nor could he by himself submit [to] the Spirit.*"

The Celtic story affirms that this destruction was caused, basically, by an asteroid, and coincides with countless remote accounts of many cultures, including the Aztec itself, since it maintains that the following era - in which mostly divine beings or immortals -, Nahui-Quiahuitl, was devastated by fire from heaven. Why was man not created directly? According to the Popol-Vuh (Mayan records), an attempt was made to create man for the first time, but there was no success, but the subsequent trial did work. This highlights a large part of other stories, implying that the very "gods" who worked on genetics and life, although they knew what they wanted to do - since the human seemed to be a prototype already existing in the universe - did not know how to do so, and it took them some time to get their ideal human, doing different biological and embryonic tests with the species that gave the impression of having the most similarities, at least on Earth: primates.

The texts of Nag Hammadi, especially, as well as the Veda records of India, declare that man already existed before coming to this world, but he lived in other worlds and in other states of "vibration", "consciousness" or "dimensionality". ", for which he needed a body as close as possible to the benefits to experience this "physicality", which the already existing forms of life did not offer him; therefore the "gods" sought to create a biological organism as competent as possible to be a vehicle for this immortal soul (man) who wished to come from these other worlds. The Sumerian novel 'Atra-Hasis' recounts that superhuman beings worked hard on earth, "*Their task was considerable, Their work was heavy, their labor infinite.*"

These were of a race called the 'Anunnaki' (referred to in the Old Testament as "the children of Anak"), but they belonged to a subgroup called the 'Igigi', or 'igigu', who obeyed the orders of the higher Anunnaki at the head of the heir. Enlil: «*[And these gods] [...] [The Igigu] (had) to excavate [the watercourses], [And open the channels] that vivify the earth. [Thus, they opened] the course of the Tigris, [and] later, [that of the Euphrates].*" *Their hard work exhausted them,* "*[For a thousand [...] years] they gave themselves to the task – [After having accumulated [...]] all the mountains, [They counted the years] worked. [After having organized [...]] the great southern swamp, [They counted [the years] worked, [(During) two thousand fi]ve hundred years, and more, They had, day and night, Endured [this heavy load!*»

They, tired of this situation, began to complain and criticize their leaders, conspiring and ending up carrying out an attempted coup, attacking the seat of power at that time. It was 'Zu' (the great sage) who infiltrated among them seemed to motivate this, according to Hittite records (which add that Zu tried to make himself a throne in the heights and be similar to the Most High), and curiously in the 'Treatise on the Origin of the World ', from Nag Hammadi, in the same order of events, appears the 'Instructor', through whom the model is designed to create man on this sphere.

They seized the mission control center and surrounded Enlil's mansion: "[*Your pa]lace is sur[rounded], my Lord! The combat[has]extended to your doorstep! – Your palace is surrounded, oh Enlil! [...] Enlil ordered the weapons to be brought to his house, - Then he opened his mouth And addressed Nuska, his page: "Nuska, build a barricade before your door! - Take your weapons and put yourself at my command!» Faced with this threat, they recommend that Enlil call his father Anu, who had decided to remain in orbit, or somewhere outside the Earth, in a celestial abode, and "Anu, the king of [Heaven], presided over (the meeting), And the king of the Apsu, Enki, listened to [everything (?)], While the great Anun[naku] [sat], Enlil stood up: the debate a[briefed]."*

This urgent assembly was organized to establish the guidelines that were going to be carried out, since the igigu refused to continue carrying this work. They voiced their complaints, while Enki, Enlil's brother, backed them up and recommended a subterfuge: "*But there is [a remedy for this situation (?)]: Since [Belet-ili, the Matrix], is here, Let me make a prot[otype of man]: He will be the one who bears the yoke [of the gods (?)]- [Whoever bears the [y]oke [of the Igigu (?)]: [Will be the Man who loads] with his [work]!»* Belet-ili assures that he would do it with the help of Enki, since he was an expert in these subjects - which we could call "genetic" -, and he affirms that then a god must be sacrificed, and his blood must be mixed with clay to make a human design.

These independent accounts relate the combination of the gods' own essence with a predesigned or existing pattern to produce the human to be created: "*The archons assembled in assembly and said: 'Come, let us land and create a man of mud". And they modeled their creature making it entirely out of dirt.*" (The Reality of Powers 1:6-7, Nag Hammadi). In another treatise from this library it is said that the authorities "ejaculated their sperm" in the "navel of the Earth" to create life, and it gives the guidelines of how this development was carried out until reaching homo sapiens sapiens. Since there would be too many sources to cite and a lot to write, we will try to summarize all these points as best as possible, adding here that everything clearly points to the use of genetic engineering to create the human being, even speaking of the use of "elements of the earth" combined for the biological manufacture of a

transport (body, avatar), but that in the first instance did not work, although theoretically everything was fine.
The body seemed "alive", but it lacked a soul (crew member), which they explain as a state of trance, coma or 'Tardema' (deep sleep). These attempts were not satisfactory from the beginning, and there were different attempts to create the "ideal" man, but when they did, they saw a danger and "mortalized" him again. Regarding the first process of creation, Enoch wrote: «*And I gave him 7 qualities, the 7 for the flesh: sight for the eyes; the smell for the soul; the joy for the ligaments-tendons; taste for blood, patience-tolerance to bones; meekness-tranquility for thought. And I thought why I said-call word cunning-wise because of what exists that is not seen and is seen I made man, both dead and living, and the image-figure knows-knows Word and there is no in all Creation like him: small in greatness and great in littleness. And I placed him as the second guardian angel on Earth, honored-upright and great, and respected-revered. And I seated him as king of the Earth and there was no one like him among my wise-cunning ones...*
... And there was no equal to him on Earth of all my creations. And I named it the 4 winds: from the East, from the West, from the North [and] from the South. And I put him in custody-appointed 4 stars of the cooled-dry ones and called his name Adam. And I gave-showed the will-dispositions and he looked at the two paths: light and darkness, and I told them: That is going well and that is bad to go, [I must] know if he has cunning-discernment towards me with hatred towards the direction excellent, [and who] of his seed loves me. And I saw my project but he did not know about the project, and if he knows he will do it badly, sin and all that is his will be sin, and I said: after sin there is no Word but death. [...] And I took the last letter of his name and called his name, they were him: Adam, [and] they [are] humanity and the living." (2nd Enoch 30:11-16) Cold stars? This was a direct way of saying "planet." Originally man in our solar system lived on 4 planets?

Did he name the four winds? The initials of the four directions in ancient Greek were 'A', 'D', 'A' and 'M'. And, what is that letter that you took for the name of the companion? Javah (Eve) was defined as "am kal-jai" (mother of all life), so her name Javah derives from Jai (life), and means "giver of life", and also, the first and last letter of the three that make up the word 'ADM' (Adam, man) form 'A.M' (mother). It was logical that she would be a mother if she were a woman, because even the rest of the creatures procreated and gave birth to their fellow men, so what this allusion intends to show is that Javah would give birth to stately beings, it would be a womb for the conception of a glorious race that embodies souls from the kingdom of heaven.

Knowing that this would happen was why "*Ialdabaot told the authorities who were with him: "Come, let us create a human being in the image of God and with the likeness of ourselves, so that this image may give us light.*" (Secret Book of John 9:1) Ialdabaot knew that the lordly man had a spark superior to him and wanted to possess it, but upon seeing his true potential he repented and destroyed this initial mold (body) producing a new mortal one, and instead to dominate everything, was dominated by the authorities to serve them and be so busy with trades and tasks that he never woke up to understand his true nature. This entire pattern was created in the image of the beings of the higher realms that Ialdabaot and his ilk had once beheld: they created with their powers and copied the features that had appeared. Each of the authorities contributed a psychic trait corresponding to the figure in the image they had seen.

In the version of Genesis, all this must be deduced based on meticulous analysis, starting with understanding that the first created human was made in the "image and likeness" of the "deity", with the purpose of ruling over all things: "*Let us make the man in our image, according to our likeness; and rule over the fish of the sea, over the birds of the heavens, over the beasts, over all the earth, and over every animal that creeps on the earth.*" (Gen. 1:26) The first design was androgynous: "*male and female he created them*" (vers. 27). However, this vegan man (verse 29) belonged to Yom Six, long before modern man, and Moses also gave evidence of this when he wrote: «*Then the heavens and the earth were finished, and all the host of them. And God finished on the seventh day the work that he had done; and he rested on the seventh day from all his work which he had done.*" (Gen. 2:1-2)

What does this mean? That when the time for cessation of this project arrived, a period - Seventh - was imposed so that things would pass by themselves, like the seed that has already been sown and watered, now only expected to germinate and produce over time. Then, after speaking of the end of all this, and of the cycle of Yom Seven, he mentions the new man who was now to be created: «*Then Yaheveh Elohim formed man from the dust of the ground, and breathed into his nostrils the breath of life, and the man was a living being.*>> (Gen. 2:7) John's version says that when creating the human, "*for a long time his creation did not move or shake at all*", and that Wisdom, to recover her power, sent 5 angels from the eternal realms to deceive Ialdabaot so that he breathed his power into the immobile creature, ergo "*He breathed his spirit into Adam*", "*Thus, the power of the Mother left Ialdabaot and entered the psychic body that had been made as the One who is from the beginning. The body moved, and became powerful. And it was enlightened.*" (Secret Book of John 10:7-8)

As we see, the first human that Moses spoke of was made the "image and likeness" of the gods, and was androgynous, but the one who came in the subsequent era was no longer produced immortal, but mortal (he had the external image of an angel, but the biological and internal resemblance of a mammal): "*from the dust of the earth*". This one, who is understood to have already had an independent sex, was deprived of immortality, but the story also seems to have a contradiction, either in the thesis or in the chronology, since much later is when a human is extracted from it. rib to make its counterpart: "*And of the rib that Yahweh Elohim took from the man, he made a woman, and brought her to the man.*" (Gen. 2:22 KJV 60)

The Nag Hammadi texts and Celtic records also mention this fact: the Egyptian Christian source maintains that this was the production of a human race with the capacity to awaken consciousness, where the woman who appeared was an image of the daughter of 'Wisdom', and this was an incarnation of this celestial woman, who awakened the man, in a sense of awareness of himself as a glorious and divine creature, letting him know that he was an incarnated soul, and that he truly came from higher realms. That is why Jesus said to John: "*Immediately the rest of the powers were jealous. Although Adam had been born through all of them, and they had given their power to this human, Adam was nonetheless more intelligent than the creators and the first ruler. When they realized that Adam was enlightened and could think more clearly than they, and was free from evil, they seized Adam and threw him to the bottom of the entire material realm.*" (Secret Book of John 10:9-10)

The Kolbrin recounts it by saying: «*Then the mist gradually cleared and the man saw another emerging form. It was that of a woman, but one such as Fanvar had never seen before, beautiful beyond his conception of beauty, with such perfection of form and grace that he was stupefied. However, the vision was not important, it was a ghost, an ethereal being.*» According to this story, this Adam, called Fanvar, was on the edge of the eastern garden and saw this ethereal entity constantly, as it deliberately appeared to him. The story says that one day the bothas (a savage yosling race (intelligent semi-human beings deprived of the capacity for spiritual transcendence)) surrounded him while he slept, and in a fierce battle he managed to get rid of them, not without receiving a slash in the face.

His side that was bleeding him, "*He became light, falling into a deep sleep and while he slept something wonderful happened: The ghost came and lay down beside [him], taking the blood from his wound to herself so that froze for her. Thus, the spiritual creature became clothed in flesh, born of coagulated blood, and being divided from his side, a mortal woman arose. In her heart Fanvar was not at rest, because of her likeness, but she was gentle, ministering to him with solicitude, and being skilled in the ways of healing, she did [to] him all. Therefore, when she had grown strong again, he made her queen of Gardenland, and she was called so, even by our fathers who named her Gulah, but Fanvar called [her] Aruah, which means 'mate'. In our language she is called the Lady of Lanevid."*

Another story, the Gilgamesh epic, tells us: «*When Anu had heard their complaints, The great Aruru they called: "You, Aruru, created man; Now create your double; With his stormy heart make him compete. Fight each other, so that Uruk may know peace!" When Aruru heard this, A double of Anu within her conceived. Aruru washed her hands, Picked up clay and threw it into the steppe. In the steppe he created the brave Enkidu, Scion of..., essence of Ninurta. Coarse hair is his whole body, He has hair on his head like a woman. The curls of her hair sprout like Nisabal.*» The human, named Adapa, had resulted from the first Mu (proto-men created by Enki to replace the work of the igigu), and although he knew the celestial abode, they deceived him so as not to reach immortality; now, Adapa was duplicated to create a being similar to him, but with whom he had to compete, Enkidu.

This very hairy human was basically a caveman, but Gilgamesh wanted to wake him up from his lethargy or absence of consciousness, so he brought a woman to him so that she would make him understand. He understood that she was flesh of his flesh and bone of his bones and he identified her as similar to him, and he was never the same again, even becoming Gilgamesh's best friend. It is possible that this story has an analogy with John's story about the last human: "*Thus, Adam became a mortal human being, the first to descend and be set apart. The enlightened Afterthought within Adam, however, would rejuvenate Adam's mind," adding that "The first ruler [...] cast oblivion upon Adam. [...] Thus said the first ruler through the prophet: "I will make their minds slow, so that they cannot understand or discern."*» (Secret Book of John 11:20-23)

Shalom!

Don't miss out!

Visit the website below and you can sign up to receive emails whenever Frederick Guttmann publishes a new book. There's no charge and no obligation.

https://books2read.com/r/B-A-DKUGB-VJHBD

BOOKS 2 READ

Connecting independent readers to independent writers.

About the Author

Israeli writer, researcher, disseminator, documentary filmmaker and influencer. He is the writer of more than 35 books, mostly research and dissemination theses.

Read more at https://www.frederickguttmann.com.

www.ingramcontent.com/pod-product-compliance
Lightning Source LLC
LaVergne TN
LVHW050344210926
848481LV00017B/1347

* 9 7 9 8 2 3 1 8 5 4 2 7 1 *